Miniature and Micro-Actuators

Guest Editor

Jose Luis Sanchez-Rojas

MDPI

Basel • Beijing • Wuhan • Barcelona • Belgrade • Novi Sad • Cluj • Manchester

Guest Editor
Jose Luis Sanchez-Rojas
Microsystems, Actuators and Sensors Lab
INAMOL-Universidad de Castilla-La Mancha
Toledo
Spain

Editorial Office
MDPI AG
Grosspeteranlage 5
4052 Basel, Switzerland

This is a reprint of the Special Issue, published open access by the journal *Actuators* (ISSN 2076-0825), freely accessible at: https://www.mdpi.com/journal/actuators/special_issues/micro_actuators.

For citation purposes, cite each article independently as indicated on the article page online and as indicated below:

Lastname, A.A.; Lastname, B.B. Article Title. *Journal Name* **Year**, *Volume Number*, Page Range.

ISBN 978-3-7258-3683-3 (Hbk)
ISBN 978-3-7258-3684-0 (PDF)
https://doi.org/10.3390/books978-3-7258-3684-0

Contents

actuators

Article

A Paper-Based Cantilever Beam Mini Actuator Using Hygro-Thermal Response

Laura Alejandra Ireta-Muñoz, Isaias Cueva-Perez, Juan Jose Saucedo-Dorantes and Angel Pérez-Cruz *

Facultad de Ingeniería, Universidad Autónoma de Querétaro, Cerro de las Campanas s/n Querétaro, Querétaro 76000, Mexico; laura.iret@gmail.com (L.A.I.-M.); isaias.cueva@uaq.mx (I.C.-P.); juan.saucedo@uaq.mx (J.J.S.-D.)
* Correspondence: angel.perez@uaq.mx

Abstract: New technological and scientific advances in the development of sensors and actuators demand the development of new devices to deal with recent problems and challenges in these new and emerging processes. Moreover, paper-based devices have tremendous potential for developing actuators as paper exhibits capillary transport and hygroexpansion due to swelling of the fibers when absorbing water. Therefore, this paper proposes a mini actuator that is based on a hygro-thermal-paper-based cantilever beam that is activated by means of a droplet of an aqueous solution in combination with a circulating electrical current to analyze its response. The contribution of this proposal includes the analysis of the flexural response of the mini actuator when it is tested by using two different solutions: distilled water and a water/alcohol solution. Additionally, four cases related to the droplet volume are studied and a statistical analysis of the bending responses is presented. The results achieved show that that water-alcohol solutions have a lower deviation in comparison with water only. Moreover, it is demonstrated that a specific change in the maximum displacement is obtained according to the volume and the type of solution. Thus, it is suggested that the response of the mini actuator can be tuned using different aqueous solutions.

Keywords: mini actuator; paper-based sensor; hygroexpansion

Citation: Ireta-Muñoz, L.A.; Cueva-Perez, I.; Saucedo-Dorantes, J.J.; Pérez-Cruz, A. A Paper-Based Cantilever Beam Mini Actuator Using Hygro-Thermal Response. *Actuators* **2022**, *11*, 94. https://doi.org/10.3390/act11030094

Academic Editor: José Luis Sanchez-Rojas

Received: 28 January 2022
Accepted: 9 March 2022
Published: 17 March 2022

1. Introduction

Currently, new technological and scientific advances are leading to complex processes in different fields; additionally, emerging technology has led to new challenges that must be addressed. On the other side, the increasing demands of large deformation devices such as wearable sensors have also led to the use of different materials in which flexibility and softness properties have been of interest. Additionally, the flexibility of soft materials has been exploited, aiming to implemented and/or apply and evaluate it in soft pneumatic actuators (SPAs) fabricated with an injection of air and silicone rubber coating [1]. Accordingly, magnetic actuators have also been implemented with soft materials to control locomotion in robots [2–4]. In this regard, the electromechanical response of soft materials has been evaluated using polymer composites to associate the actuator displacement with the changes in dielectric constant [5]. Particularly, thermal actuators have been implemented using the bending response caused by the application of an electric field which is related to the speed of actuator response, aiming to provide solutions for recent applications [6]. However, challenges have recently arisen in the development of new actuators; the fabrication process requires a long time or specialized equipment to fabricate such devices. Moreover, the material is non-biodegradable. For this reason, the investigation of biodegradable materials has gained importance in the development of actuators [7]. On the other hand, paper and graphene have been used and combined to contribute to the development of strain sensors [8]. In this sense, the flexibility of these materials has been also exploited for the development of soft-actuators and humidity sensors [9]. Likewise, other properties such as electrical resistance have been also analyzed in order to provide feasible routes for constructing efficient sensors that may produce efficient electrical stimuli [10]. Yet although recent studies have contributed and lead to significative advances, a drawback has arisen as this material requires specialized treatment for its deposition on paper.

Paper-based microfluidic systems have been broadly studied due to the characteristics associated with the acronym ASSURED (Affordable, Sensitive, Specific, User-Friendly Rapid and Robust, Equipment Free, Deliverable to all end-users) [11]. These characteristics have been exploited and have led to the development of diverse systems such as lab-on-a-chip devices in applications such as the photoelectrochemical (PEC) visual sensing systems [12,13]; additionally, they have been implemented in analytical devices that represent affordable solutions, capable of measuring microfluidic responses [14,15], and such characteristics have intentionally led to the design and development of sophisticated devices that have been implemented as a part of personalized health care procedures including emergency, ambulatory, and remote areas [16]. Additionally, it should be highlighted that properties such as biodegradability and low cost make paper a suitable option for the development of gas, humidity, strain, and pressure sensors [17–19]. Other application fields are involved with the development of sensors in medical applications, such as using paper for physiological monitoring [20], where properties such as the elasticity of paper have been the object of study for the design and fabrication of wearable devices [21]. Furthermore, the flexibility of paper has also been considered for the implementation of wearable devices in health monitoring [22,23].

Therefore, the aforementioned properties, such as biodegradability and flexibility, have been studied to evaluate the response of paper to different actuation stimuli [24]. One of the most important properties is its mechanical response which has been studied and addressed through the analysis of the response under tension and compression loads [25]. These properties have been implemented together with the electric response of the material to develop electromechanical actuators [26]. Electroactive paper (EAPap), which bends when an electric field is applied to the electrodes of the device [27], has actuation principles that have been implemented using different combinations of materials [28]. Thermal actuation has also been used to measure bending angles in a bonding beam made of graphite paper and polyimide film (PI) when an electric field is applied [29]. Furthermore, the response to temperature gradients has also been implemented for the design of electric switches [30]. The motion in paper-based systems has been also studied under the effects of fluid-related phenomena such as capillarity [31] and moisture gradients that cause curling on paper [32]. Relative and local humidity changes have also been associated with the mechanical response of paper bending by hygroexpansion [33] and have been combined with the use of light irradiation [34]. Hamedi et al. at [35] implemented a hygroexpansive electrothermal paper actuator (HEPA) in a bilayer configuration which consists of paper, conducting polymer, and adhesive tape. This device uses a conductive path to generate the necessary heat, and in combination with the principle of absorption of moisture causes actuation. Moreover, the influence of humidity and temperature changes on the mechanical response of paper-based actuators has been studied using numeric methods [36]. These humidity-based actuators have a faster response to humidity changes and a good response to the thermal effect generated by the conductive path, but the control of humidity was implemented in a controlled chamber with a humidifier. Nevertheless, the use of a localized moisture gradient caused by a droplet has not been implemented to induce actuator movement. This actuation principle may be of interest in the field as it can be fully integrated with paper-based microfluidic devices.

Therefore, this paper proposes a hygro-thermal-mechanical paper-based mini actuator triggered by a microliter droplet of an aqueous solution. The mini actuator consists of a cantilever beam with a conductive path to accelerate the evaporation time of aqueous solutions. To evaluate the flexural actuator response, different solution volumes were tested in the actuator. The local increment of moisture content on the mini actuator is produced by a droplet of an aqueous solution placed directly on the surface of the mini actuator with a micropipette. The movement is induced by the hygroexpansion effect, and the 'going-back' motion through the evaporation which is accelerated by a current across the conductive path. Two aqueous solutions were used to evaluate the response of the actuator: distilled water, used as a reference case, and a mix of distilled water with alcohol using different volumes. It was found that the maximum deflection produced by the actuation principle is associated with the solution volume and that the time response to reverse this motion is associated with the solution's characteristics. The results show that distilled water with alcohol produces a faster actuation in comparison with distilled water due to the faster evaporation of alcohol.

2. Materials and Methods

In this section the working principle, materials, and experimental set up are described. Likewise, the proposed mini actuator and tuning response by droplet volumes are presented.

2.1. Mini Actuator Working Principle

The matrix of cellulose-based materials provides an excellent medium to transport liquids without external forces. When paper interacts with liquids, its mechanical properties are modified, i.e., by the softening of paper and the reduction in its elasticity module [37–39]. This is caused by the interaction between the liquid and the cellulose fibers. The cellulose fibers are expanded or contracted due to this interaction. As a result, deformation phenomena such as creeping, waving, curling, and deflection are observed at the macroscopic level [40]. Perez et al. [41] took advantage of this phenomenon by developing a paper-based device for the characterization of binary aqueous solutions. This was achieved by modelling the bending response of a paper-based cantilever beam. However, this response is in the static/quasi-static domain due to the slowness of the deflection response. The increasing demand of eco-friendly, affordable, and portable devices involving paper-based mechanical systems demands more knowledge about the dynamic behavior of humidified/wetted paper to produce paper-based actuators with faster responses.

The proposed mini actuator is based on the hygro-thermal-mechanical response of paper that consists of a cantilever beam with a conductive path that was fabricated to accelerate the evaporation time of the aqueous solutions. Its working principle is depicted following its activation response, which is induced by means of the imbibition of an aqueous solution into the cantilever beam paper. The activation is illustrated precisely in Figure 1. First, Figure 1a shows the proposed mini actuator in its initial position, where the left end is the fixed end, which is connected to a current supply. Afterwards, a droplet of an aqueous solution is deposited onto the surface of the cantilever beam, as can be seen in Figure 1b; subsequently, the fibers of paper swell as their moisture content increases—this is namely hygroexpansive strain. It has been suggested that swelling takes place as the water molecules break and replace interchain bonds in cellulose [41]. As an example, Figure 2 shows the hygroexpansive strain of dry and wet chromatography paper. In this case, the hygroexpansive strain of 0.17 is measured along the thickness of the paper. Thus, the motion or the activation response is produced due to the differential expansion of paper and the conductive layer; this is represented in Figure 1c. As the thermal linear expansion of the silver conductive path is in the order of 10^{-3} while the hygroexpansion is two orders of magnitude greater (10^{-1}), the contribution of thermal expansion is neglected in this work. Finally, aiming to reduce the deactivation time (deswelling of fibers) of the actuator due to drying process of paper, the heating produced by the conductive electrode incorporated on the top of the beam leads the actuator to return to its initial position once is fully dried, as depicted in Figure 1d. The conductive path has been designed to reach a maximum temperature of approximately 40 °C to not compromise the integrity of paper but provide heating to dry the mini actuator once is wetted after activation, following Ansari and Cho [42].

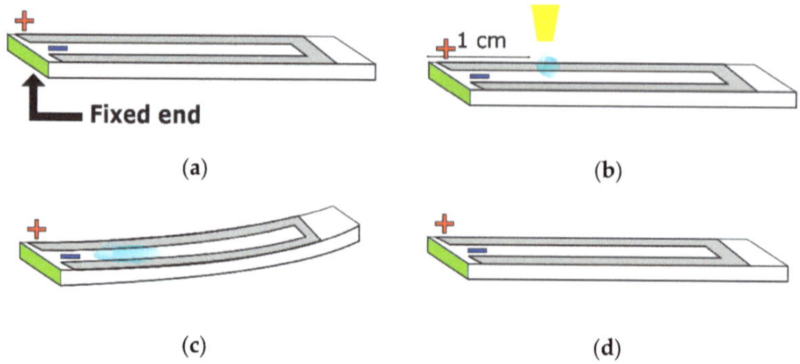

Figure 1. Schematic of the actuation principle: (**a**) Initial position; (**b**) Activation by volume solution; (**c**) Movement induced by hygroexpansion; (**d**) Deactivation/deswelling induced by current heating; return to initial position.

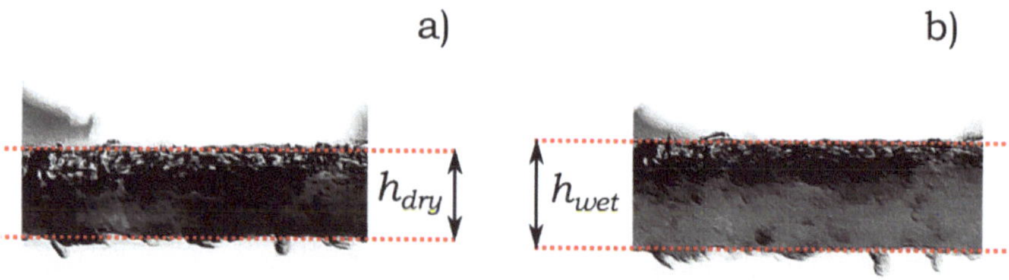

Figure 2. Microscope cross section image of a piece of chromatography paper experiencing hygroexpansive strain along its thickness. Comparison of (**a**) dry thickness (hdry) and (**b**) wet thickness (hwet).

The reaction produced in the cantilever beam is quantified measuring the displacement at the free end of the actuator; the whole activation procedure is shown in Figure 3. In this sense, the mini actuator has an initial position as described in Figure 3a; this initial position belongs to pinpoint (a) in the graph at initial time. Subsequently, the mini actuator remains connected to the control current for five minutes. After this period, a droplet solution is added to the beam as shown in Figure 3b; afterward, the paper hygroexpansion induces a movement until a maximum displacement is reached; this located in the graph as pinpoint (c) and is depicted by Figure 3c. Finally, due to the temperature reached due to the conductive path, the actuator tip returns to a point that is close to its initial position.

2.2. Experimental Set Up

To evaluate the performance of the proposed paper-based mini actuator, different cantilever beams were designed with similar geometrical characteristics such as a length, width, and thickness of 35 mm, 5 mm, and 1 mm, respectively. All the actuators are made of filter CST paper from Triton Electronics Ltd., Dunmow, Essex, UK. The silver conductive ink is hand-printed on top of the cantilever beam; silver was selected because of its good performance on porous materials like paper and because it remains over the top layer and is not imbibed into the paper fibers. The shape and dimensions of the conductive path are selected to ensure a specific resistance value of 2.9 ohms. Thus, the geometrical representation of the proposed paper-based mini actuator, as well as the materials employed, are presented in Figure 4a; it can be observed that the conductive path is located on top of the filter paper and has a width of 1 mm. The beams are cut out manually using a conventional knife, trying to retain as much similarity as possible. Thus, the final size of the cantilever beam may be compared to the real size of a two-euro coin, which is portrayed in Figure 4b.

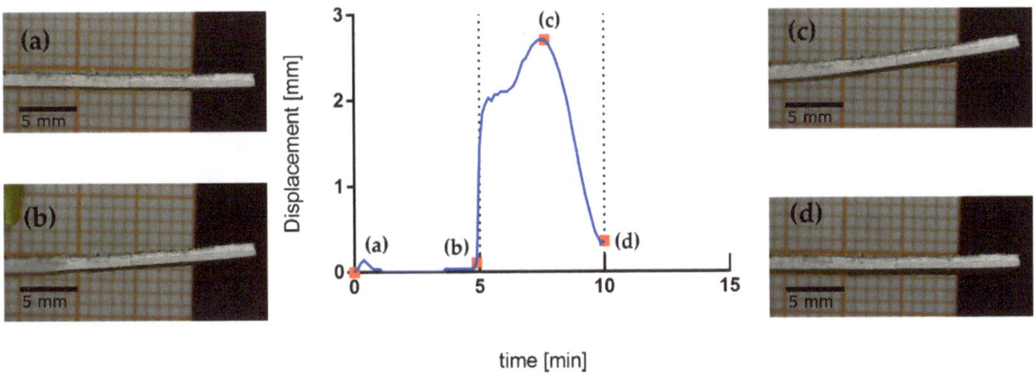

Figure 3. Activation/deactivation of paper cantilever-beam: (**a**) Initial position when connected to current control; (**b**) After 5 min, activation by volume solution; (**c**) Maximum displacement induced by hygroexpansion; (**d**) Deactivation/evaporation induced by current; return to initial position.

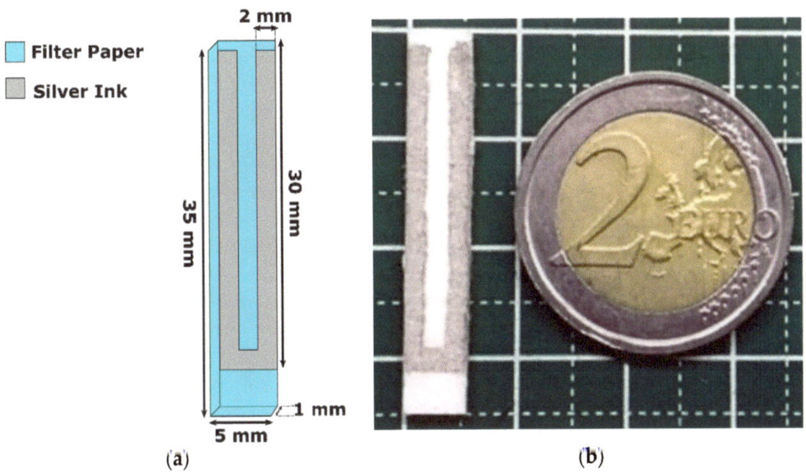

Figure 4. Comparative geometry schematic for sample mini actuator (**a**) Geometry and materials of mini actuator; (**b**) Representation of the silver conductive ink hand printed on top of the cantilever beam and its size compared with a two euro coin.

The experiment was performed in a controlled environment under laboratory conditions; in Figure 5, the complete experimental test bench is shown, and consists of a box of Plexiglas, with a holder to fix one of the ends of the cantilever beam (Figure 5A). In order to ensure the repeatability of the experiments, the experimental test bench also includes a support for the pipette and an illumination system in order to take pictures for post-processing purposes (Figure 5B). To capture and measure the deflection response of the cantilever beam, several pictures were acquired by handheld microscope and stored in a personal computer (Figure 5C). Each tested beam was electrically connected to current control circuitry to heat the conductive path during the experiment (Figure 5D). For further details on control circuity please refer to Supplementary Figure S1. As can be observed in Figure 5D, the current control was connected to a mini actuator using a connector. Finally, the circuitry was connected to a regulated power supply designed to deliver 5V and to feed the illumination system (Figure 5E).

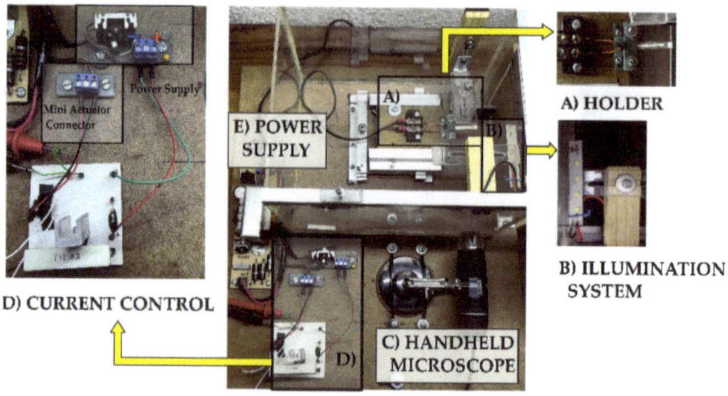

Figure 5. Experimental setup used for testing and to validate the proposed mini actuator composed by: (**A**) the holder, (**B**) the illumination system, (**C**) the handheld microscope, (**D**) the current control and (**E**) the power supply.

2.3. *Tuning Response of the Mini Actuator*

Aiming to tune the response of the actuator produced by a distilled water droplet (20 µL), the actuator response was first evaluated without the application of current. The obtained response can be observed in Figure 6, where the red curve represents the actuator response without current application; as can be observed, the actuator produced a flexural response where the bending activation was faster than deactivation. Such a phenomenon is due to the deactivation procedure taking a long time, since there is no current flow in the conductor path and the temperature of the actuator cannot be increased; consequently, the paper fibers take a long time to dry out. On the other side, also in Figure 6, the blue curve depicts the obtained response related to the flexural displacement when several drops of water are deposited in the actuator and when its conductive path is connected to current control for adding heat. Due to the actuator path being subjected to current excitation, an increase in the actuator temperature is achieved, allowing a faster dry-out; as can be observed in Figure 6, the return to the initial position becomes faster with each droplet of solution deposited. The current values are measured when the actuator path is subjected to the current flow; thus, Figure 7 shows the current values measured during the experiment each activation. It is possible to observe that variation in current values remains in a range of 385 to 458 mA, which ensures the maximum temperature calculated.

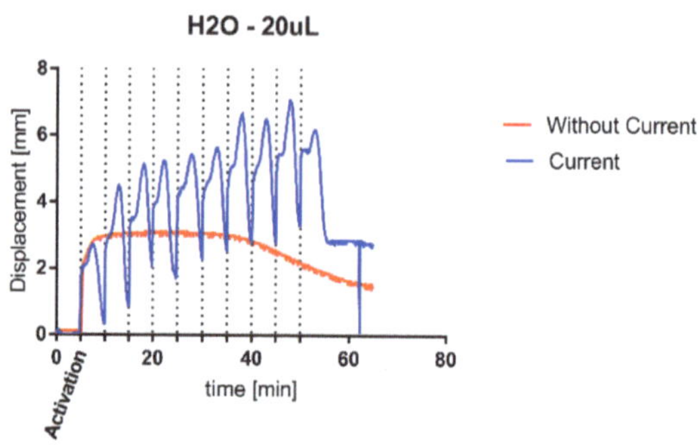

Figure 6. Activation/deactivation time of paper-beam for 20 µL distilled water.

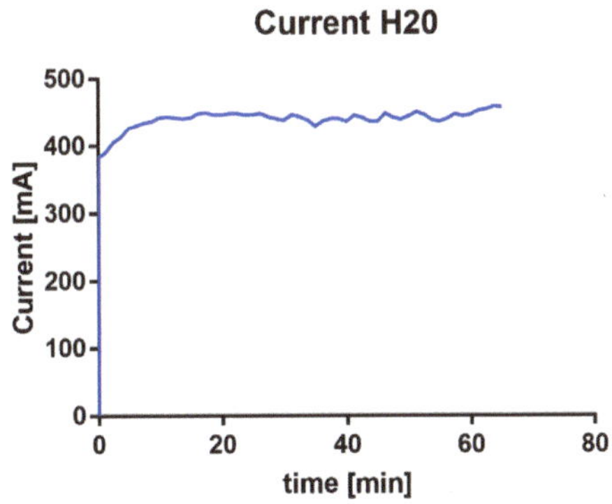

Figure 7. Current measured during experiment with 20 µL of aqueous solution.

Although the conductive path is set to reach a maximum of 40 °C, it is expected that the wetting activation of the paper-based actuator has a key role on the thermal response of the beam. In this case, the authors proposed two means of tuning the response of the mini actuator: (i) controlling the amount of heat while using water, and (ii) using alternative aqueous solutions while setting a constant current throughout the conductive path. The latter is adopted in this work; thus, a binary solution of distilled water and alcohol is suggested for tuning the mini actuator response as its evaporation rate is faster than that of distilled water. Moreover, a tuning of the bending response of the mini actuator by varying the volume of water-alcohol is proposed. It is expected that, with a decrease in the volume of solution deposited, the amplitude of deflection would decrease, for two reasons. First, as the volume of liquid is decreased, a small hygroexpansion would occur. Second, with a low volume of droplets deposited on the beam, the time for drying would be reduced.

In this regard, the displacement measurement was performed for seven samples per volume deposited, which consisted of 20 µL of distilled water, and 20 µL, 15 µL, and 10 µL of distilled water with alcohol in a proportion 50/50 *v/v*. The proposed volume of the solution was selected by trial and error with the purpose of assessing the actuator response. Due to its evaporation rate, the actuator response should return to the initial position faster than in the distilled water case. The droplet was placed between the larger edges of the beam, at 1 cm of length measured from the fixed end of the sample, as was shown previously in Figure 2b. Subsequently, this experiment consisted of supplying a current level to the mini actuator for 65 min to obtain ten activations per each tested cantilever beam. The movement activation in the beam was performed by adding a droplet of solution to the sample every 5 min; this time was selected because it was observed that it represents the mean time that a sample requires to return to the reference point. Subsequently, pictures of the beam were handled by a microscope, which was configured with a sampling time of 7.4 s; the pictures were processed with the software ImageJ to obtain the displacement change in the sample. In Figure 8a, a representation of the reference point for an initial time is shown. Figure 8b illustrates the maximum deflection at the free end of the beam and the reference level (dashed yellow line).

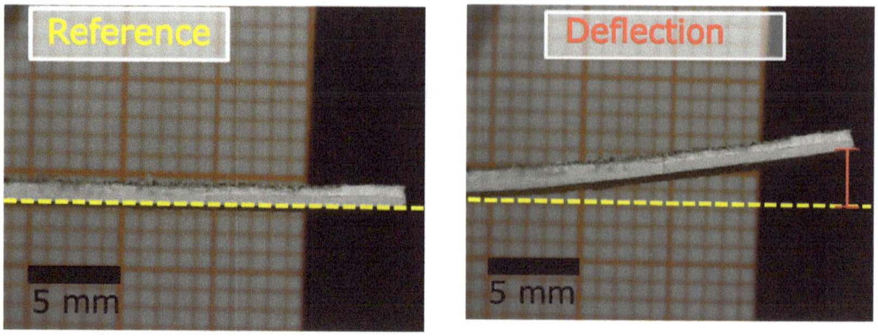

(a) (b)

Figure 8. Deflection measurement of the free end of the sample to measure the displacement changes with respect to the reference line: (**a**) Reference guideline; (**b**) Measure deflection.

3. Results and Discussions

The results achieved support and validate the proposed mini actuator and are presented in this section as well as the details and discussion regarding each experiment performed. In order to obtain statistically significant results, a set of ten activations in a row were performed on each mini actuator to evaluate its performance. Thus, in Figure 9 the displacements per beam for each corresponding amount of solution are represented graphically. As can be observed, the achieved displacements are reported in millimeters; in particular, the first case corresponds to a solution of 20 μL of distilled water [H_2O] as is shown in Figure 9a. Hence, Figure 9a shows the behavior of the seven beams and the activation obtained by the addition of a micro droplet of water every five minutes. Moreover, as can be observed, the measured displacement presents an increase that is associated with the activation of each droplet solution by the addition of a micro droplet of water every five minutes. During the experimentation, it was noticed that for some cases the displacement of the beam rarely returned to the reference point. This phenomenon was present because the water was absorbed by cellulose fibers due to hygroexpansion. For that reason, the response obtained and shown in Figure 9a presents the curve tendency for each sample displacement increased with time without reaching the reference point.

Accordingly, the remained of the experiments were carried out using a binary solution composed of distilled water-alcohol. The alcohol used in this experiment was ethanol, which produces a faster drying process in the paper due to its higher volatility, in comparison with the volatility of water. It is well known that volatility depends on the strength of the liquid intermolecular bonds; in this regard, water has stronger intermolecular bonds in comparison with ethanol. Therefore, it is easier for ethanol to break the interface between liquid and gas when kinetic energy is added to the liquid. In this case, the kinetic energy is added through the Joule effect caused by the current circulating in the conductive path. Thus, it can be concluded that solutions in which water is mixed with ethanol have a faster drying process. In this way, the second case also considers a solution volume of 20 μL of distilled water-alcohol; the solution was applied as previously described in Figure 9b, where the achieved increases in displacement are shown. As can be appreciated, the resulting tendency shows behavior that is contrary to the first case of study (Figure 9a); specifically, the reference point is reached by each activation. This is because the evaporation time is faster due to the chemical properties of alcohol. Likewise, Figure 9c,d show the obtained displacements that belong to the third and fourth cases of study, that is, for 15 μL and 10 μL, respectively. As can be observed in Figure 9c,d, the maximum displacement reached by each activation is smaller than in the two previous cases. Additionally, the same behavior is observed in the third and fourth cases of study. Through these obtained results, it can be concluded that the maximum displacement is associated with fiber expansion due to the solution volume.

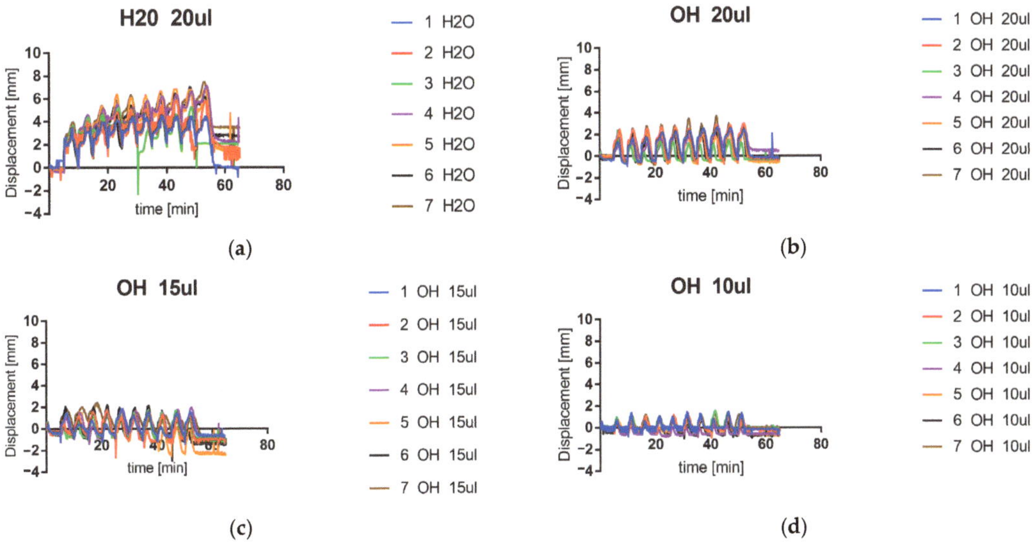

Figure 9. Measured displacement of each sample per amount of solution: (**a**) 20 µL distilled water; (**b**) 20 µL distilled water with alcohol (OH); (**c**) 15 µL distilled water with alcohol (OH); (**d**) 10 µL distilled water with alcohol (OH).

Furthermore, the experimental results show that the displacement per sample is different for each type and volume solution. For this reason, in order to complement the achieved results, a statistical analysis was performed by estimating the maximum and mean/average displacement. Thus, the average displacement and the maximum displacement were first computed for each analyzed sample; afterwards, the average values per each case of study were obtained for both the mean displacement and the maximum displacement, averaging the results obtained for each sample. Such comparison was performed among the studied samples, estimating the standard deviation. As a result, the maximum displacement average and the standard deviation for each experiment are shown in Figure 10, where the horizontal axis corresponds to each study case and the vertical axis represents the mean value. The error bar corresponds to one standard deviation. Specifically, in Figure 10a the results obtained for maximum displacement are shown; here, a tendency for the displacement to increase is observed. The values reached for the binary solutions were 1.144 mm, 1.917 mm, and 2.827 mm for the cases of 10 µL, 15 µL, and 20 µL, respectively. On the other side, for the test with 20 µL of distilled water a maximum average value of 6.340 mm was obtained, which is the highest value of all the cases evaluated. Here, it can be noted that the use of a binary solution not only reduces the time response but also the hygroexpansion of paper. The error bar for the case of water is greater than the other cases; this may be related to the evaporation rate and the large deformations of the beam due to the high expansion of the fibers. In Figure 10b the mean values for each case of study are shown; as can be observed, the tendency is not clear due to the quantity of values computed and slight differences between each case.

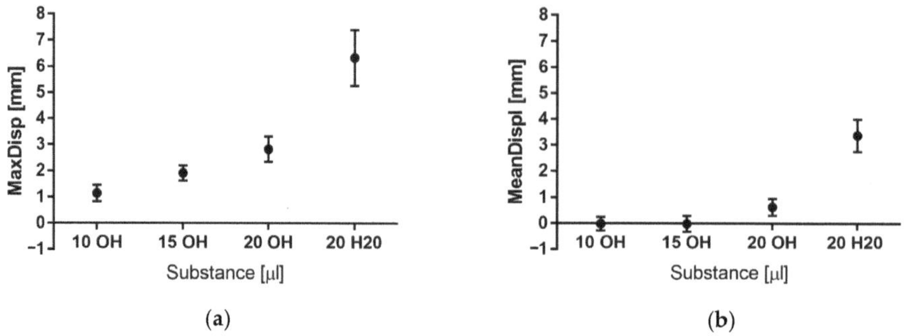

Figure 10. Mean values per study case: (**a**) Maximum displacement; (**b**) Mean displacement.

4. Conclusions

In this work, a hygro-thermal-mechanical paper-based mini actuator triggered by a microliter droplet of an aqueous solution was proposed. The novelty of this proposal considers the design of the mini actuator to be installed as a cantilever beam with a conductive path that leads to accelerated evaporation times of the aqueous solutions. The proposed mini actuator is validated by its evaluation through its flexural response with different solution volumes. In this regard, the response of the mini actuator was evaluated for four cases. The first case study was carried out with distilled water, using 20 μL of solution. The following three cases corresponded to the binary solution of alcohol and distilled water. The amount of liquid placed on the samples in each experiment was 10, 15, and 20 μL. From the results of these studies, it was concluded that the maximum displacement measured for each case was associated with the amount and type of solution. Thus, it is possible to tune the response of the mini actuators using other aqueous solutions. Additionally, the results suggest that the proposed mini actuator is suitable for microfluidic applications, as it is activated by liquid transport, and its application in different fields may lead to the solution of recent challenges/drawbacks that cannot be addressed by conventional actuators. An automated calibration procedure based on embedded circuit technology could be developed in future works, as well as numerical modeling of the thermal and wetting phenomena to improve the design and to ensure the repeatability of the device, with regard to the displacement levels that were reached in this present work. Furthermore, the development of the proposed mini actuator should lead to new devices that can be used in several applications in the field of microfluidics—certainly, in applications such as electrochemical biosensors and piezoelectric biosensors.

Supplementary Materials: The following supporting information can be downloaded at: https://www.mdpi.com/article/10.3390/act11030094/s1, Figure S1: Current Control Diagram.

Author Contributions: Conceptualization, A.P.-C. and L.A.I.-M.; methodology, A.P.-C. and J.J.S.-D.; validation, I.C.-P. and L.A.I.-M.; formal analysis, I.C.-P. and L.A.I.-M.; writing—original draft preparation, L.A.I.-M. and I.C.-P.; writing—review and editing, A.P.-C. and J.J.S.-D. All authors have read and agreed to the published version of the manuscript.

Funding: This research was partially supported by the National Council on Science and Technology of Mexico (CONACYT) through the scholarship 312230.

Informed Consent Statement: Not applicable.

Data Availability Statement: Data available on request.

Acknowledgments: The authors acknowledge the technical support for the experimental set up provided by Fernando Ireta Moreno.

Conflicts of Interest: The authors declare no conflict of interest.

References

1. Oguntosin, V.; Abdulkareem, A. Design of a pneumatic soft actuator controlled via eye tracking and detection. *Heliyon* **2020**, *6*, e04388. [CrossRef]
2. Venkiteswaran, V.K.; Tan, D.K.; Misra, S. Tandem actuation of legged locomotion and grasping manipulation in soft robots using magnetic fields. *Extreme Mech. Lett.* **2020**, *41*, 101023. [CrossRef]
3. Tang, D.; Zhang, C.; Sun, H.; Dai, H.; Xie, J.; Fu, J.; Zhao, P. Origami-inspired magnetic-driven soft actuators with programmable designs and multiple applications. *Nano Energy* **2021**, *89*, 106424. [CrossRef]
4. Fu, C.; Xia, Z.; Hurren, C.; Nilghaz, A.; Wang, X. Textiles in soft robots: Current progress and future trends. *Biosens. Bioelectron.* **2022**, *196*, 113690. [CrossRef]
5. Kirkman, S.; Rothemund, P.; Acome, E.; Keplinger, C. Electromechanics of planar HASEL actuators. *Extreme Mech. Lett.* **2021**, *48*, 101408. [CrossRef]
6. Zhao, G.; Yang, J.; Wang, Y.; Zhao, H.; Wang, Z. Preparation and electromechanical properties of the chitosan gel polymer actuator based on heat treating. *Sens. Actuators A Phys.* **2018**, *279*, 481–492. [CrossRef]
7. Migliorini, L.; Santaniello, T.; Rondinini, S.; Saettone, P.; Comes Franchini, M.; Lenardi, C.; Milani, P. Bioplastic electromechanical actuators based on biodegradable poly(3-hydroxybutyrate) and cluster-assembled gold electrodes. *Sens. Actuators B Chem.* **2019**, *286*, 230–236. [CrossRef]
8. Qi, X.; Li, X.; Jo, H.; Sideeq Bhat, K.; Kim, S.; An, J.; Kang, J.; Lim, S. Mulberry paper-based graphene strain sensor for wearable electronics with high mechanical strength. *Sens. Actuators Phys.* **2020**, *301*, 111697. [CrossRef]
9. Jia, G.; Zheng, A.; Wang, X.; Zhang, L.; Li, L.; Li, C.; Zhang, Y.; Cao, L. Flexible, biocompatible and highly conductive MXene-graphene oxide film for smart actuator and humidity sensor. *Sens. Actuators B Chem.* **2021**, *346*, 130507. [CrossRef]
10. Wang, W.; Liu, D.; Liu, Y.; Leng, J.; Bhattacharyya, D. Electrical actuation properties of reduced graphene oxide paper/epoxy-based shape memory composites. *Compos. Sci. Technol.* **2015**, *106*, 20–24. [CrossRef]
11. Smith, S.; Korvink, J.G.; Mager, D.; Land, K. The potential of paper-based diagnostics to meet the ASSURED criteria. *RSC Adv.* **2018**, *8*, 34012–34034. [CrossRef]
12. Zhang, L.; Kong, Q.; Li, L.; Wang, Y.; Ge, S.; Yu, J. Direct-readout photoelectrochemical lab-on-paper biosensing platform based on coupled electricity generating system and paper supercapacitors. *Talanta* **2021**, *222*, 121517. [CrossRef] [PubMed]
13. Kong, Q.; Cui, K.; Zhang, L.; Wang, Y.; Sun, J.; Ge, S.; Zhang, Y.; Yu, J. "on-Off-On" Photoelectrochemical/Visual Lab-on-Paper Sensing via Signal Amplification of CdS Quantum Dots@Leaf-Shape ZnO and Quenching of Au-Modified Prism-Anchored Octahedral CeO_2 Nanoparticles. *Anal. Chem.* **2018**, *90*, 11297–11304. [CrossRef] [PubMed]
14. Ozer, T.; McMahon, C.; Henry, C.S. Advances in Paper-Based Analytical Devices. *Annu. Rev. Anal. Chem.* **2020**, *13*, 85–109. [CrossRef] [PubMed]
15. Lim, W.Y.; Goh, B.T.; Khor, S.M. Microfluidic paper-based analytical devices for potential use in quantitative and direct detection of disease biomarkers in clinical analysis. *J. Chromatogr. B Analyt. Technol. Biomed. Life. Sci.* **2017**, *1060*, 424–442. [CrossRef] [PubMed]
16. Mahato, K.; Srivastava, A.; Chandra, P. Paper based diagnostics for personalized health care: Emerging technologies and commercial aspects. *Biosens. Bioelectron.* **2017**, *96*, 246–259. [CrossRef] [PubMed]
17. Tai, H.; Duan, Z.; Wang, Y.; Wang, S.; Jiang, Y. Paper-Based Sensors for Gas, Humidity, and Strain Detections: A Review. *ACS Appl. Mater. Interfaces* **2020**, *12*, 31037–31053. [CrossRef] [PubMed]
18. Long, Y.; He, P.; Xu, R.; Hayasaka, T.; Shao, Z.; Zhong, J.; Lin, L. Molybdenum-carbide-graphene composites for paper-based strain and acoustic pressure sensors. *Carbon* **2020**, *157*, 594–601. [CrossRef]
19. Lee, D.; Kim, D. Paper-Based, Hand-Painted Strain Sensor Based on ITO Nanoparticle Channels for Human Motion Monitoring. *IEEE Access* **2019**, *7*, 77200–77207. [CrossRef]
20. Huang, Y.; Wang, Z.; Zhou, H.; Guo, X.; Zhang, Y.; Wang, Y.; Liu, P.; Liu, C.; Ma, Y.; Zhang, Y. Highly sensitive pressure sensor based on structurally modified tissue paper for human physiological activity monitoring. *J. Appl. Polym. Sci.* **2020**, *137*, 48973. [CrossRef]
21. Liu, H.; Qing, H.; Li, Z.; Han, Y.L.; Lin, M.; Yang, H.; Li, A.; Lu, T.J.; Li, F.; Xu, F. Paper: A promising material for human-friendly functional wearable electronics. *Mater. Sci. Eng. R Rep.* **2017**, *112*, 1–22. [CrossRef]
22. Liu, Y.; Wang, H.; Zhao, W.; Zhang, M.; Qin, H.; Xie, Y. Flexible, Stretchable Sensors for Wearable Health Monitoring: Sensing Mechanisms, Materials, Fabrication Strategies and Features. *Sensors* **2018**, *18*, 645. [CrossRef] [PubMed]
23. Wang, X.; Liu, Z.; Zhang, T. Flexible Sensing Electronics for Wearable/Attachable Health Monitoring. *Small* **2017**, *13*, 1602790. [CrossRef] [PubMed]
24. Liu, Y.; Shang, S.; Mo, S.; Wang, P.; Yin, B.; Wei, J. Soft actuators built from cellulose paper: A review on actuation, material, fabrication, and applications. *J. Sci. Adv. Mater. Devices* **2021**, *6*, 321–337. [CrossRef]
25. Li, Y.; Stapleton, S.E.; Reese, S.; Simon, J.W. Anisotropic elastic-plastic deformation of paper: Out-of-plane model. *Int. J. Solids Struct.* **2018**, *130–131*, 172–182. [CrossRef]
26. Kanaparthi, S.; Sekhar, V.R.; Badhulika, S. Flexible, eco-friendly and highly sensitive paper antenna based electromechanical sensor for wireless human motion detection and structural health monitoring. *Extreme Mech. Lett.* **2016**, *9*, 324–330. [CrossRef]
27. Kim, J. Chapter 7—Nanocellulose-based paper actuators. In *Nanocellulose Based Composites for Electronics*; Thomas, S., Pottathara, Y.B., Eds.; Elsevier: Amsterdam, The Netherlands, 2021; pp. 163–183. [CrossRef]

28. Khan, A.; Abas, Z.; Kim, H.; Kim, J. Recent Progress on Cellulose-Based Electro-Active Paper, Its Hybrid Nanocomposites and Applications. *Sensors* **2016**, *16*, 1172. [CrossRef] [PubMed]
29. Fan, Q.; Miao, J.; Tian, M.; Zhao, H.; Zhu, S.; Liu, X.; Ma, Y.; Qu, L. Low-voltage Driven Flexible Double-layer Electrothermal Actuator for Smart Human-machine Interactions. *Sens. Actuators Phys.* **2020**, *315*, 112352. [CrossRef]
30. Shin, S.; So, H. Time-dependent motion of 3D-printed soft thermal actuators for switch application in electric circuits. *Addit. Manuf.* **2021**, *39*, 101893. [CrossRef]
31. Ha, J.; Kim, H.Y. Capillarity in Soft Porous Solids. *Annu. Rev. Fluid Mech.* **2020**, *52*, 263–284. [CrossRef]
32. Brandberg, A.; Motamedian, H.R.; Kulachenko, A.; Hirn, U. The role of the fiber and the bond in the hygroexpansion and curl of thin freely dried paper sheets. *Int. J. Solids Struct.* **2020**, *193–194*, 302–313. [CrossRef]
33. Kumar, A.; Heidari-Bafroui, H.; Charbaji, A.; Rahmani, N.; Anagnostopoulos, C.; Faghri, M. Numerical and Experimental Modeling of Paper-Based Actuators. *Chem. Proc.* **2021**, *5*, 15. [CrossRef]
34. Weng, M.; Tang, Z.; Zhu, J. Multi-responsive soft paper-based actuators with programmable shape-deformations. *Sens. Actuators Phys.* **2021**, *331*, 113016. [CrossRef]
35. Hamedi, M.M.; Campbell, V.E.; Rothemund, P.; Güder, F.; Christodouleas, D.C.; Bloch, J.; Whitesides, G.M. Electrically Activated Paper Actuators. *Adv. Funct. Mater.* **2016**, *26*, 2446–2453. [CrossRef]
36. Zhou, L.; Nie, B.; Ren, S.; Żur, K.K.; Kim, J. On the hygro-thermo-electro-mechanical coupling effect on static and dynamic responses of piezoelectric beams. *Compos. Struct.* **2021**, *259*, 113248. [CrossRef]
37. Lee, M.; Kim, S.; Kim, H.Y.; Mahadevan, L. Bending and buckling of wet paper. *Phys. Fluids* **2016**, *28*, 042101. [CrossRef]
38. Douezan, S.; Wyart, M.; Brochard-Wyart, F.; Cuvelier, D. Curling instability induced by swelling. *Soft Matter.* **2011**, *7*, 1506–1511. [CrossRef]
39. Reyssat, E.; Mahadevan, L. How wet paper curls. *Europhys. Lett.* **2011**, *93*, 54001. [CrossRef]
40. Cueva-Perez, I.; Perez-Cruz, A.; Stiharu, I.; Dominguez-Gonzalez, A.; Trejo-Hernandez, M.; Osornio-Rios, R.A. A parametric study on the nonlinear dynamic response of paper-based mechanical systems due to liquid transport. *Int. J. Non-Linear Mech.* **2020**, *118*, 103280. [CrossRef]
41. Perez-Cruz, A.; Stiharu, I.; Dominguez-Gonzalez, A. Nonlinear imbibition influence on the hygro-mechanical bending response of paper due to its interaction with water. *Int. J. Non-Linear Mech.* **2017**, *97*, 89–95. [CrossRef]
42. Ansari, M.Z.; Cho, C. A conduction–convection model for self-heating in piezoresistive microcantilever biosensors. *Sens. Actuators A Phys.* **2012**, *175*, 19–27. [CrossRef]

actuators

MDPI

Article

3D-Printed Miniature Robots with Piezoelectric Actuation for Locomotion and Steering Maneuverability Applications

Víctor Ruiz-Díez *, José Luis García-Caraballo, Jorge Hernando-García and José Luis Sánchez-Rojas

Microsystems, Actuators and Sensors Lab, School of Industrial Engineering, Universidad de Castilla-La Mancha, E-13071 Ciudad Real, Spain; joseluis.gcaraballo@uclm.es (J.L.G.-C.); jorge.hernando@uclm.es (J.H.-G.); joseluis.saldavero@uclm.es (J.L.S.-R.)
* Correspondence: victor.ruiz@uclm.es

Abstract: The miniaturization of robots with locomotion abilities is a challenge of significant technological impact in many applications where large-scale robots have physical or cost restrictions. Access to hostile environments, improving microfabrication processes, or advanced instrumentation are examples of their potential use. Here, we propose a miniature 20 mm long sub-gram robot with piezoelectric actuation whose direction of motion can be controlled. A differential drive approach was implemented in an H-shaped 3D-printed motor platform featuring two plate resonators linked at their center, with built-in legs. The locomotion was driven by the generation of standing waves on each plate by means of piezoelectric patches excited with burst signals. The control of the motion trajectory of the robot, either translation or rotation, was attained by adjusting the parameters of the actuation signals such as the applied voltage, the number of applied cycles, or the driving frequency. The robot demonstrated locomotion in bidirectional straight paths as long as 65 mm at 2 mm/s speed with a voltage amplitude of only 10 V, and forward and backward precise steps as low as 1 μm. The spinning of the robot could be controlled with turns as low as 0.013 deg. and angular speeds as high as 3 deg./s under the same conditions. The proposed device was able to describe complex trajectories of more than 160 mm, while carrying 70 times its own weight.

Keywords: robot; standing wave; piezoelectric; leg; miniature; steering; 3D printing

Citation: Ruiz-Díez, V.; García-Caraballo, J.L.; Hernando-García, J.; Sánchez-Rojas, J.L. 3D-Printed Miniature Robots with Piezoelectric Actuation for Locomotion and Steering Maneuverability Applications. *Actuators* **2021**, *10*, 335. https://doi.org/10.3390/act10120335

Academic Editor: Kenji Uchino

Received: 12 November 2021
Accepted: 15 December 2021
Published: 20 December 2021

1. Introduction

Miniaturization of mobile robots is a growing field of research that has attracted increased attention over many years, despite the remaining challenges to overcome. The first generation of micro-robots was on the rise in the late 1980s and the early 1990s, targeting cm-sized, self-propelling, and climbing machines. Many industries and academic researchers have invested in the study of the principle of motion of various insects, in what could be considered the original period of 'bio-inspired' engineering [1]. However, the development of micro-machines and micro-robots is still a current topic of great interest. A recent article by St. Pierre et al. [2] highlighted the difficulties associated with the decrease in size of the robots concerning speed, control, and autonomy. The achievement of such miniature-sized robots would imply lower cost and accessibility to areas forbidden to larger robots [3].

The fabrication of miniature sub-gram robots is a remarkable feat, where different fabrication technologies and actuation techniques have demonstrated encouraging results [4–7]. Alternatively, 3D printing is a promising approach that allows for the manufacture of 3D objects by successive layer-by-layer deposition with high resolution and flexibility in geometry and materials [8–11]. When dealing with miniature robots, the minimum feature size of the 3D printing technique is a major determinant of the accuracy of the design. Besides, the stiffness of the material to be printed is a critical parameter to consider that affects the performance of the device from a mechanical point of view. Among the different

3D printing techniques, laser-based printers offer enough resolution when pursuing milli sized robots [10], and a photopolymer with a tensile modulus up to 10 GPa after curing is already commercially available [12].

Regarding the actuation mechanism of small size robots, a wavelike motion of the body of the robot, with attached passive legs, is a well stablished approach. The combination of waves of standing nature (i.e., a vibration mode) with legs was already proposed in the seminal paper by He et al. [13] for slider motion. Recent articles have reported the achievement of standing-wave (SW) locomotion of robots at the mm-sized scale [14,15]. A step further is to steer the robot. Recent reports have shown the turning of milli-sized robots by only tuning the frequency of actuation of the device [11,16]. More complex steering techniques rely on individual control of the legs [17,18] or a differential-drive approach by independent control of each side of the robot [19–21].

The long-term target of our work is the development of a low-cost sub-gram autonomous robot capable of running through complex paths with high accuracy and requiring low voltage signals. The robot should be able to carry power, processing, communication, and sensing devices, with a good balance between control and agility. Several engineering challenges needed to be overcome to reach this goal, beginning with the design and manufacture of a sub-gram low-cost locomotion platform. In this regard, here, we propose a miniature 20 mm long sub-gram robot with piezoelectric actuation whose direction of motion can be controlled. The fabrication of the robots was based on 3D printing and the locomotion of the printed robots was controlled by the generation of standing waves (SW) associated with resonant bending modes. Various printing materials were considered but stiffer materials resulted in a better performance in terms of quality factor and conductance of the generated SWs. Regarding the steering technique, a differential drive approach was implemented by means of two legged plates, joined together at the middle length (H-shaped). The generation of SWs at each of the plates, together with the proper location of the legs, allowed for the bidirectional locomotion. The actuation was achieved by piezoelectric patches and the control of the motion path was attained by controlling parameters such as the frequency of actuation, the applied voltage, or the number of applied cycles (duty cycle). In terms of performance, the proper control of the signals applied to the plates allowed us to achieve a straight path at 2 mm/s speed, and to perform forward and backward reproducible steps with a resolution as low as 1 µm. In addition, the spinning of the robot could be controlled with turns as low as 0.013 deg. and angular speeds of 3 deg./s. With these optimizations, the proposed design demonstrated the accomplishment of complex trajectories of more than 160 mm while carrying a payload 70 times its own weight.

2. Device Design

Figure 1 shows the type of structure under consideration. The body of the robot consisted of a pair of rectangular plates with a length of $L_s = 20$ mm, width of $W_s = 3$ mm, and thickness of $t_s = 0.85$ mm, bound together by a light nexus of the same material. This nexus consists of a rectangular section bridge, with lengths (L_{nx}) varying from 1 to 3 mm. The actuation of the robot was carried out by lead zirconate titanate (PZT) piezoelectric patches that started at one edge and covered a given length (L_p) on each plate to be determined by design.

The locomotion of the robot was based on the generation of standing waves (SW) in the two plate resonators, in combination with an appropriate location of legs [13]. This configuration was successfully applied to the bidirectional linear motion of millimeter-sized robots by means of two consecutive flexural modes [15,22]. The two plates of our design could be considered as independent SW motors and were designed as such.

Here, we considered the flexural modes with five and six nodal lines along the length of the plate and no nodal lines along the width of the plate. Using a two-index naming convention representing the number of nodal lines along the length and the width, we will refer to these as the (50) and (60) modes. Figure 2 shows a 1D representation of the considered mode shapes. In order to optimally excite either of these modes, the piezoelectric patch should cover the area at which the curvature of the mode shapes has a constant sign [15,22]. In our case, in the search of the ease of fabrication, a patch length of 5 mm on both plates allowed for an efficient actuation of both modes.

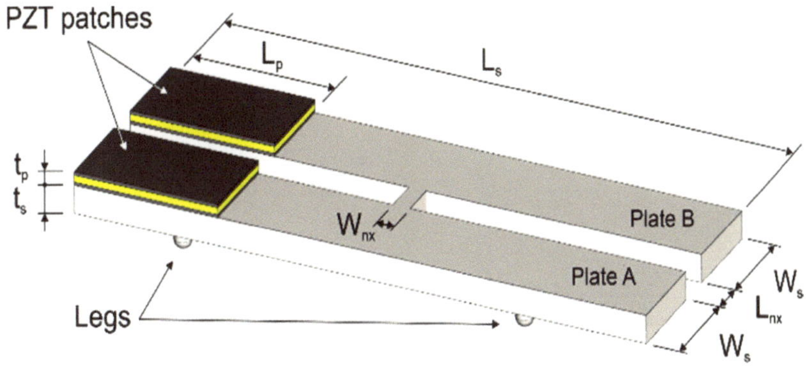

Figure 1. Schematic description of the twin-plate robot design.

The location of the legs is the key point of the SW-based bidirectional locomotion. Figure 2 shows where to locate them according to the criteria explained by He et al. [13]. We searched for positions that allowed for opposite movements for each modal vibration. Therefore, the location of the legs should be at the right of a crest (red areas) for a given modal shape, and to the left of the crest (blue areas) for a subsequent modal shape. Consequently, by tuning the frequency of actuation to the resonance frequency of these modes, the direction of the movement could be changed. The legs should be located where the blue and the red areas overlap, highlighted by the dashed rectangles in Figure 2, which results in five possible positions for locating the legs.

In search of a compromise between stability and friction losses, we decided to use a total of three legs attached to the robot: two legs at plate A, located at positions 1 and 4, and only one leg at position 3 for plate B. The size of the legs was optimized in a previous work for a similar SW-based motor [15]. A leg length of 1.1 mm and a diameter of 0.6 mm was chosen, so that the leg first resonance was much higher than the driving frequencies of the robot to prevent plate–leg coupling effects.

Our design allows for four different types of motion that can be seen in Figure 3. Each plate could be actuated on the (50) or (60) mode by applying sinusoidal signals at the corresponding resonant frequencies. By combining either of the modes on each of the two plates, four distinct movements could be obtained by differential drive: forward and backward translational movement and clockwise (CW) or counterclockwise (CCW) rotation of the robot.

Due to the asymmetries present on the structure caused by the leg distribution or the fabrication tolerances, the thrust exerted by the plates was not the same, despite applying the same signal to each one. This resulted in deviations from the ideal expected movement. To correct the trajectory, a fine-tuning of the signal applied to each of the plates was required to balance out the unwanted deviations. Both the voltage amplitude and the number of cycles of the actuation signal were chosen as the parameters for the adjustment, while the frequency of actuation was fixed to the value corresponding to the resonance frequency of the required modal shape. Burst-type signals were used, consisting of a finite number of sinusoidal cycles. The period of the burst signal, $T_b = T_{ON} + T_{OFF}$ was fixed to one second. The duration T_{ON} of the active part of the signal was equivalent to the number of cycles times the period of the excitation signals (i.e., $T_{ON} = \frac{cycles}{f_{drive}}$). This gives a duty cycle $D = \frac{T_{ON}}{T_b}100 = \frac{cycles}{f_{drive}}100$. These parameters are exemplified in Figure 4a.

Two alternative actuations and therefore balancing procedures were considered: either by voltage amplitude or by number of cycles. In the first one, depicted in Figure 4a, the same number of excitation cycles was applied on both plates and the thrust balance was achieved by adjusting the ratio of applied voltage amplitudes. In the second alternative, depicted in Figure 4b, the thrust was balanced by adjusting the number of cycles in the active part of the burst signals with the same voltage amplitude for both signals.

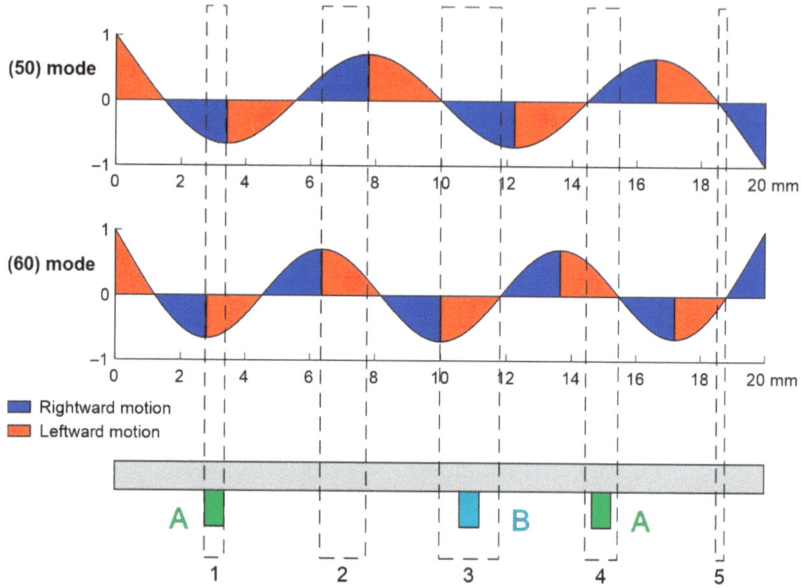

Figure 2. Schematics showing the location of the legs for standing wave (SW)-based locomotion with the (50) and (60) modes. Each colored half lobe represents the region where a leg produces a rightward or leftward movement of the robot. For bidirectional motion, legs should be located on the dashed vertical rectangles. Positions 1 and 4 were chosen for plate A and position 3 for plate B.

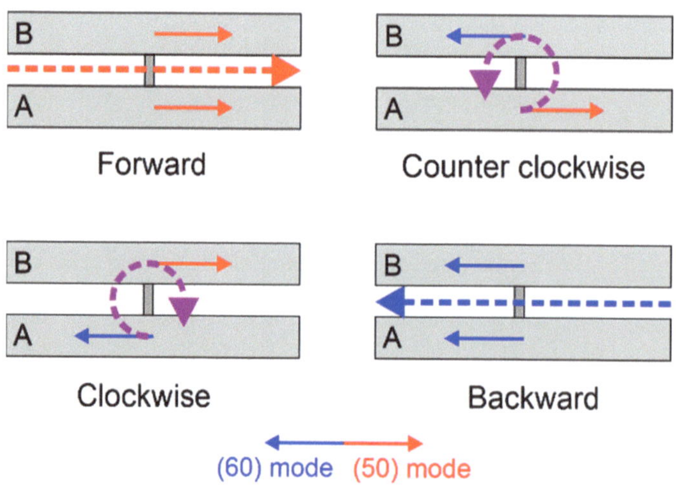

Figure 3. Graphical description of the four different types of motion of the robot. Each plate could be actuated either on the (50) or (60) mode for a bidirectional thrust.

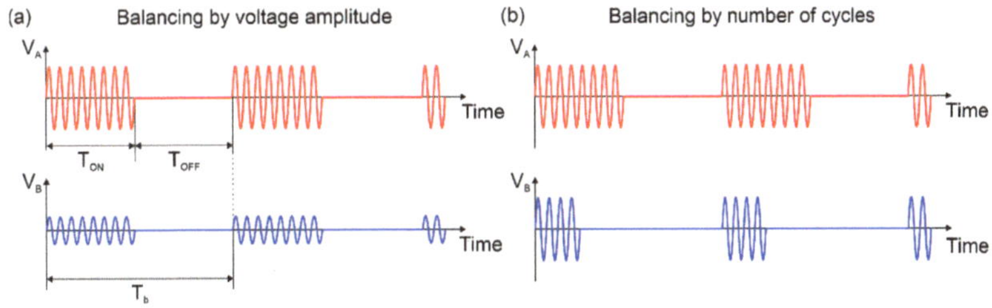

Figure 4. Excitation voltage signals applied to the plate resonators A (**red color**) and B (**blue color**). Each burst signal is characterized by the voltage amplitude, the burst period (T_b), and the active period (T_{ON}). Two alternative actuations were considered: (**a**) both plates with the same number of sinusoidal cycles per burst but different voltages, or (**b**) both plates with the same voltage amplitude but different number of cycles.

3. Materials and Methods

Following the design guidelines in the previous section, several robots were fabricated, as shown in Figure 5. The procedure is as follows.

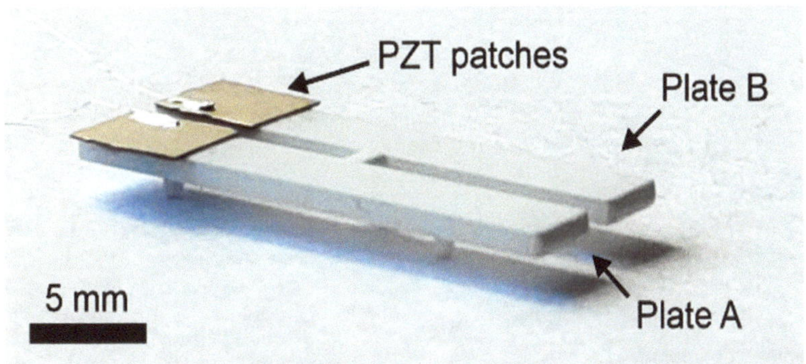

Figure 5. Photograph of one of the fabricated robots. The two legs of plate A can be observed. Electrical access to the PZT patches was made through 25 µm silver insulated wires.

First, an H-shape supporting platform consisting of two 20 mm long, 0.85 mm thick and 3 mm wide plates attached by a nexus was fabricated on a highly glass-filled resin (Formlabs Rigid 10 K resin [23]) in a Form 3 stereolithography (SLA) printer by Formlabs. Different sizes of the central nexus were considered with lengths varying from 1 to 3 mm and cross sections of 0.2×0.2 mm^2 and 0.6×0.85 mm^2.

The monolithic structure included 1.1 mm long cylindrical legs with a 0.6 mm diameter. Two legs were printed at the center of the width of plate A and an additional leg at the center of the width of plate B, at those positions along the length described in the design section. After printing, the structure was cured with ultraviolet (UV) light and heat treatment for 60 min at 70 °C, acquiring its final properties (Table 1).

Table 1. Material properties of the different components used in the robot fabrication.

Element	Thickness (mm)	Young's Modulus (GPa)	Density (kg/m^3)
PZT patches	$t_P = 0.1$	62	6700
Rigid 10K resin	$t_s = 0.85$	10	1670

Next, two electroded PZT patches (PIC 255 from PI Ceramic GmbH, Lederhose, Germany) with a thickness $t_p = 100$ µm and a width of 3.5 mm, slightly wider than the plate to allow electrical access to the bottom electrode, were glued to each printed plate by means of a cyanoacrylate adhesive (Loctite, Düsseldorf, Germany). The length of the patch was 5 mm, according to Figure 2. Additionally, 25 μm in diameter silver insulated wires were attached to the piezoelectric patches for electrical connection to the external signal generator. The mechanical properties of the PZT patches can be found in Table 1.

The kinetic characterization was carried out employing two different recording cameras. One with high speed, high resolution, and a large field of view (Logitech Streamcam) was used to track the robot while performing large displacements. This camera covered a 23×13 cm^2 area, with a spatial resolution of 120 microns at 60 frames per second (fps). In addition, a microscope camera consisting of a CMOS sensor (Throcam DCC1545M) coupled to an optic tube was used to record the position and the orientation of the robot while performing small steps. This camera covered a 2×2 mm^2, with a spatial resolution of 100 nm at 30 fps. The cameras were computer controlled as was the signal generator (Tektronix AFG 3000 series) used to apply the required sinusoidal signals to generate the SW in the plates. The robot demonstrated locomotion in a wide variety of surfaces such as glass, polyimide, polyethylene naphthalate (PEN), cast aluminum, or a flat surface made of the same material of the robot platform, Rigid 10K resin. However, the glass surface showed better results in terms of performance and trajectory control, so the experiments were all carried out with the robot lying on a leveled glass surface. Finally, the recorded videos were processed by a motion tracking algorithm programmed in MATLAB.

4. Results

Prior to the subsequent kinetic analysis, the electrical response of the resonators was characterized by means of an impedance spectrum analysis. A 4294A Agilent impedance analyzer was used to record the conductance versus frequency of the different modes of vibration on each of the two plate resonators that comprised the robots. The figures of merit of the electrical response such as the quality factor, the resonant frequency, and the peak conductance were calculated through the fitting of these impedance measurements to a modified Butterworth–Van-Dyke equivalent electrical model circuit [24]. Besides, the detected impedance peaks were associated with the corresponding flexural modes with the help of a scanning laser Doppler vibrometer (Polytec MSV 400).

Figure 6 shows the conductance spectra of one of the fabricated robots. The H-shaped platform, comprising the two plates with legs and the joint linking the plates, was printed as a monolithic part, so the obtained resonant frequencies were determined by the whole structure. As can be seen, there were two clear peaks that were identified as the (50) and (60) modes. The figures of merit of these impedance peaks are shown in Table 2. The resonant frequency of the (50) and (60) modes were 59 and 86 kHz, respectively, while similar Q-factors around 26 and motional conductance in between 65 and 117 μS were estimated. Similar results were obtained for all the fabricated samples with the same nominal dimensions of the plates. Differences in the resonant frequencies below 2% were obtained among the fabricated samples, demonstrating the reproducibility of the process.

Table 2. Calculated figures after fitting of the impedance data in Figure 6 to a modified Butterworth–Van-Dyke equivalent circuit model.

Resonant Mode	(50) Mode			(60) Mode		
Plate	Frequency (kHz)	Q-Factor	ΔG (μS)	Frequency (kHz)	Q-Factor	ΔG (μS)
A	59.3	28	65	85.8	24	79
B	59.2	27	72	86.1	25	117

Once the electrical response of the constituent plates was analyzed, the kinetic characterization of the robot was addressed. First of all, the determination of the appropriate excitation signals for a thrust balance in all four basic movements in Figure 3 was accomplished. Figure 7 shows the results corresponding to the fine tuning of amplitudes to balance out the deviations from the forward motion. As can be seen in Figure 7a, the different ratios of the applied voltage from plates A to B resulted in a range of trajectories with dominant forward motion. Figure 7b depicts the off-axis absolute deviation (Y-displacement to X-displacement ratio) of these trajectories. The ratio of voltages with the lowest deviation from the X-axis was then chosen for this value of equal excitation cycles. The same procedure was performed for the backward movement and for the two turning possibilities in Figure 3, minimizing the displacement of the robot's center of rotation, obtaining the voltage ratios for a balanced thrust in every case. Similarly, this procedure could be performed for equal amplitude of applied voltages on both plates, but different ratios of number of excitation cycles.

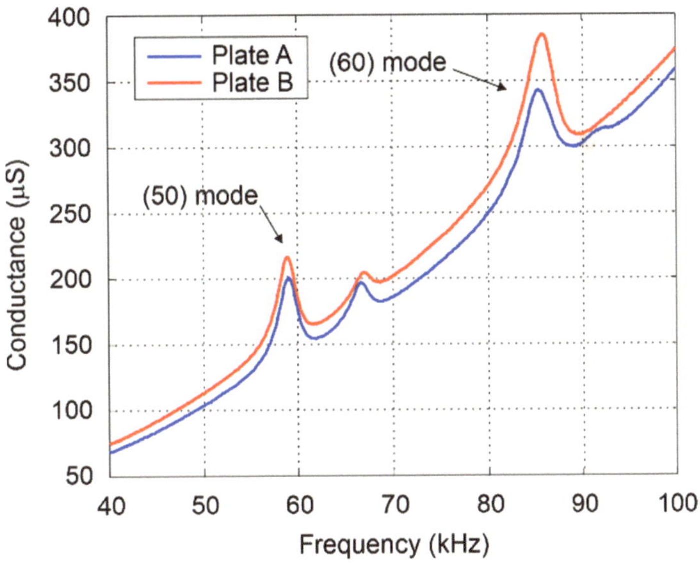

Figure 6. Conductance spectra of plates A and B of one of the fabricated robots. The main peaks were identified as the (50) and (60) modes.

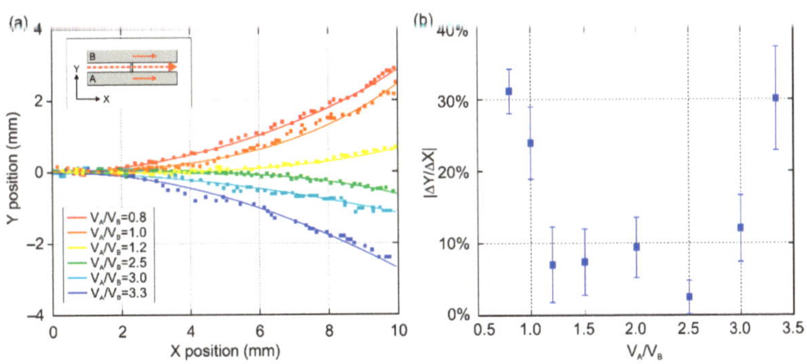

Figure 7. (**a**) Forward motion trajectories for different ratios of applied voltages on each plate. (**b**) Percentual absolute deviation off the X-axis displacement in forward motion for different ratios of applied voltages on each plate. All experiments were carried out at a constant number of excitation cycles of 7000 on both plates.

4.1. Speed Control

Figure 8 shows the results from the speed characterization for the forward and backward locomotion in Figure 3. The thrust of the plates was balanced following the above-described procedures, allowing the robot to make linear displacements up to 65 mm with a controlled deviation off the *X*-axis. The large FOV camera setup was used in these experiments.

Figure 8a,b shows the mean speed of the robot for forward motion, where both plates were excited at the (50) mode resonant frequency. The results from the balancing procedure are also depicted in Figure 8a,b, with markers representing the combinations for a balanced motion by voltage amplitude or excitation cycles, respectively. As can be seen, both procedures allowed reaching a maximum speed of about 1.85 mm/s by (i) applying 6 V and 16,000 cycles on plate A and 6V and 6000 cycles to plate B or (ii) applying 6 V and 16,000 cycles to plate A and 3 V and 16,000 cycles to plate B. Similar results were obtained for other input combinations. Therefore, for a balanced forward motion, plate A required twice as much power than plate B, either by applying a 2:1 voltage ratio at equal number of cycles or a 8:3 number of cycles ratio at equal applied voltage.

The backward motion (Figure 8c,d), where both resonators were excited at (60) mode resonant frequency, showed analogous results. The mean speed was lower when compared to the forward motion, and higher voltage amplitudes were required. Similar differences were found in bidirectional SW robots implemented in glass in [15] and might be attributed to the tight margin for the location of the legs to achieve bidirectional movement for the SW-based approach. Finally, in this case, the balancing ratios by voltage amplitude and excitation cycles were 0.6 and 0.5, respectively.

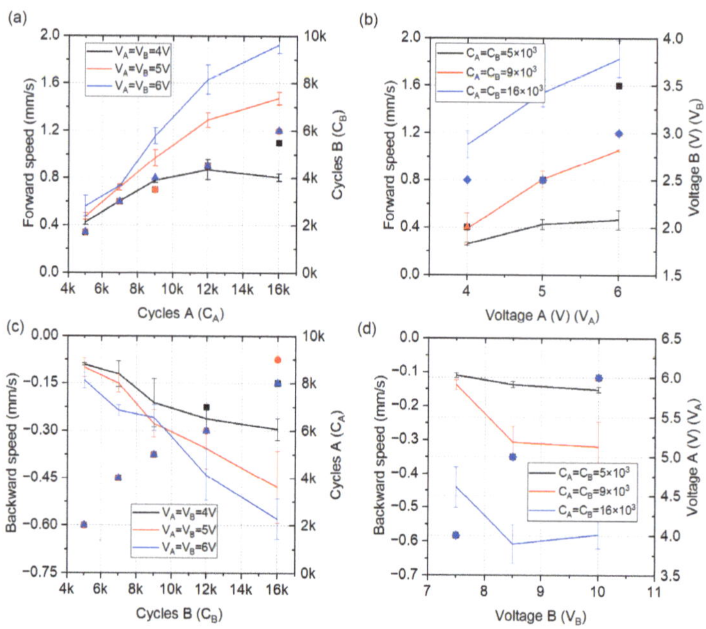

Figure 8. Results of translational speeds (solid lines and left vertical axes) in the forward and backward motions of Figure 3. Each color represents a different voltage amplitude, $V_A = V_B$, (**a**,**c**) or a different number of excitation cycles, $C_A = C_B$, (**b**,**d**) applied on both plates. The particular combinations of the number of cycles or the voltage amplitudes required for a balanced motion are represented with markers related to the right vertical axes.

The implemented differential drive allowed for CCW and CW rotations. Figure 9 shows the results from the angular speed characterization for the two turning possibilities in Figure 3. The thrust of the motors was balanced as in the previous experiments, allowing the robot to make rotations up to 360 deg., recorded with the large FOV camera setup. An angular speed of above 3 deg./s was obtained for both directions using either the same voltage (Figure 9a,c) or the same number of excitation cycles (Figure 9b,d) on both plates.

In addition, as can be inferred from the balancing results in Figure 9, regardless of the sense of rotation, the plate actuated with the (50) mode, (plate A in CCW and plate B in CW) required about three times a lower number of cycles or applied voltage. This could also be attributed to the non-ideal leg location for bidirectional SW robots [15].

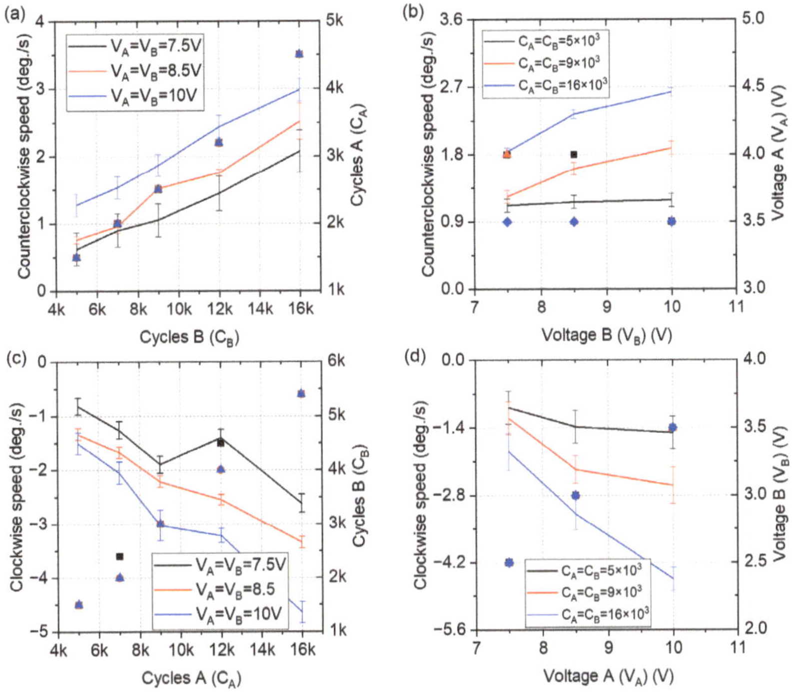

Figure 9. Results of angular speeds (solid lines and left vertical axes) in the CCW and CW motions of Figure 3. Each color represents a different voltage amplitude, $V_A = V_B$, (**a**,**c**) or a different number of excitation cycles, $C_A = C_B$, (**b**,**d**) applied on both plates. The particular combinations of the number of cycles or the voltage amplitudes required for a balanced motion are represented with markers related to the right vertical axes.

4.2. Positional Control

One of the main advantages of the actuation strategy consisting of burst signals is the capability of the robot to move in discrete steps with high resolution. We studied the positional capabilities of the robot by reducing the supplied energy of the excitation signals in search of the smallest reproducible translation or rotation. Figure 10 shows the results from the experiments where a reduced number of cycles was equally applied to both plates. In this case, the balancing was performed by applying different voltage ratios to each plate while recording the robot movement with the high-resolution recording camera setup.

Regarding translational movements, as can be seen in Figure 10a,c, the step length could be decreased to less than 1 µm. This value was close to the limit of detection of the used setup, so lower translational steps could be achievable. The forward motion balance was also reached for a 2:1 voltage ratio, as in the speed experiments (see Figure 8b), and a decreasing trend of the step size with the number of cycles or the application voltage was also observed. In contrast, the backward motion balance was reached at a 3:4 voltage ratio and required higher voltages to get to the minimum detectable step, than in forward motion. The decreasing trend with the number of excitation cycles was not so prominent when considering the applied voltage.

Figure 10b,d shows the results of the CCW and CW movements, respectively, when the number of excitation cycles were reduced to less than 2000. The minimum turn was 0.02 deg., but voltages above 6 V were required to produce detectable rotations. The balance was reached at a 2:1 voltage ratio in the CW rotations, and similar but not so consistent results were obtained for the CCW turns.

A summary of the main results from the speed and positional characteristics of the robot is shown in Table 3. The maximum speeds were taken from the results with a $T_b = 1$ s and the upper limits of the experimental ranges (10 V and 16,000 cycles) while the minimum steps were taken from the experiments with the lower input limits of each type of motion.

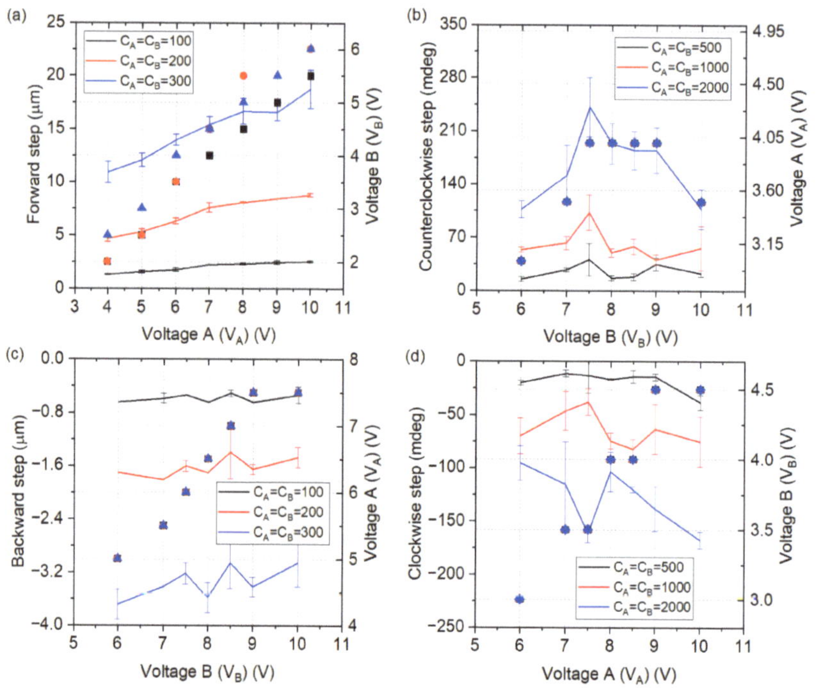

Figure 10. Results in terms of minimum translational steps and turns (solid lines related to the left vertical axes) from the positional experiments in the four different types of movement of Figure 3. Each color represents a different number of excitation cycles applied on both plates ($C_A = C_B$). For a balanced motion, different combinations of applied voltages, noted with markers and related to the right vertical axes, were required.

Table 3. Summary of the main results from the kinetic characterization of the robots.

Type of Motion	Maximum Speed	Minimum Movement
Forward	1.85 mm/s	1 µm
Backward	0.6 mm/s	0.4 µm
Counterclockwise	3 deg./s	0.016 deg.
Clockwise	4.7 deg./s	0.013 deg.

4.3. Mass Loading

The effect of mass loading on the performance of the robots was also investigated. Figure 11 shows the robot speeds for the four movements depicted in Figure 3 for different payloads and two distinct applied voltages of 5 V and 10 V using the large FOV camera setup. The excitation cycles were used to balance the motion, with values around 7000 for the dominant plate on each basic movement (right in the middle of the range used in the experiments of Figures 8 and 9). As can be seen, the robot was able to carry 15 g, 70 times its own weight, at 0.8 mm/s and 0.2 mm/s for forward and backward motion, respectively. Additionally, the robot was able to develop angular speeds of 2.5 deg./s and 1.1 deg./s for the CCW and CW directions, respectively, at the maximum applied voltage. The loading capacities of our proposed design make it suitable for autonomous applications, where the robot should carry onboard circuits including the control and the power systems, with a total weight below our achieved payload [25,26].

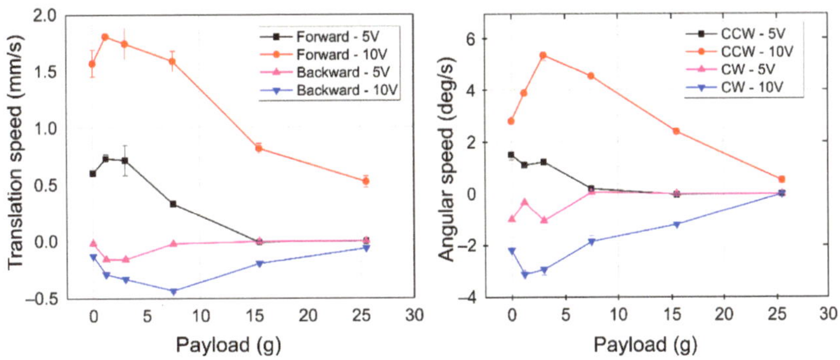

Figure 11. Translational and rotational speeds for different payloads and two applied voltages.

It is also noticeable from the results in Figure 11 that a maximum in the robot speed was found for loading masses in the range of 1 to 3 g. This maximum could be attributed to a compromise between the frictional forces, which increase with the addition of mass, and the vibration amplitude, which decreases due to the added mass damping.

In addition, Table 4 shows the blocking force of the robot under different mass loadings. The force was measured while the robot contacted a force sensor (Honeywell FSG Series) with the actuation voltage applied. A maximum blocking force of 10 mN was measured for a payload of 15 g and a continuous excitation of 10 V, while the robot was moving forward. Similar blocking force values were obtained in glass-based single-plate robots with SW actuation but for half the mass loading [15].

Table 4. Blocking force of the robot actuated on forward-direction movement for various payload and voltage conditions.

Blocking Force (mN)	**Payload (g)**					
Voltage (V)	**0**	**1**	**3**	**7.5**	**15**	**25**
5	0.3 ± 0.1	1.1 ± 0.1	2.0 ± 0.2	3.5 ± 0.1	3.7 ± 0.1	2.8 ± 0.1
7.5	0.3 ± 0.1	1.5 ± 0.1	2.4 ± 0.1	5.3 ± 0.1	7.0 ± 0.3	5.3 ± 0.4
10	0.6 ± 0.1	1.9 ± 0.1	2.8 ± 0.1	7.0 ± 0.3	10 ± 0.3	9.4 ± 0.5

4.4. Complex Trajectory

Finally, the kinetic performance, the trajectory control, and the mass loading capabilities of the proposed design were tested through a series of preprogramed sequences of movements. No real-time control strategy was used in these experiments. The robot was loaded with 15 g and actuated in such a way that a right triangular trajectory and a square path were described at maximum speed for 10 V and $T_b = 0.5$ s. The same adjustment ratios as in the previous experiments with $T_b = 1$ s were used, demonstrating the validity of the control procedure for a different burst signal period. Figure 12 shows the results of the performance of these paths, and the video experiments, recorded in the larger FOV setup, are provided in Supplementary Videos S1 and S2.

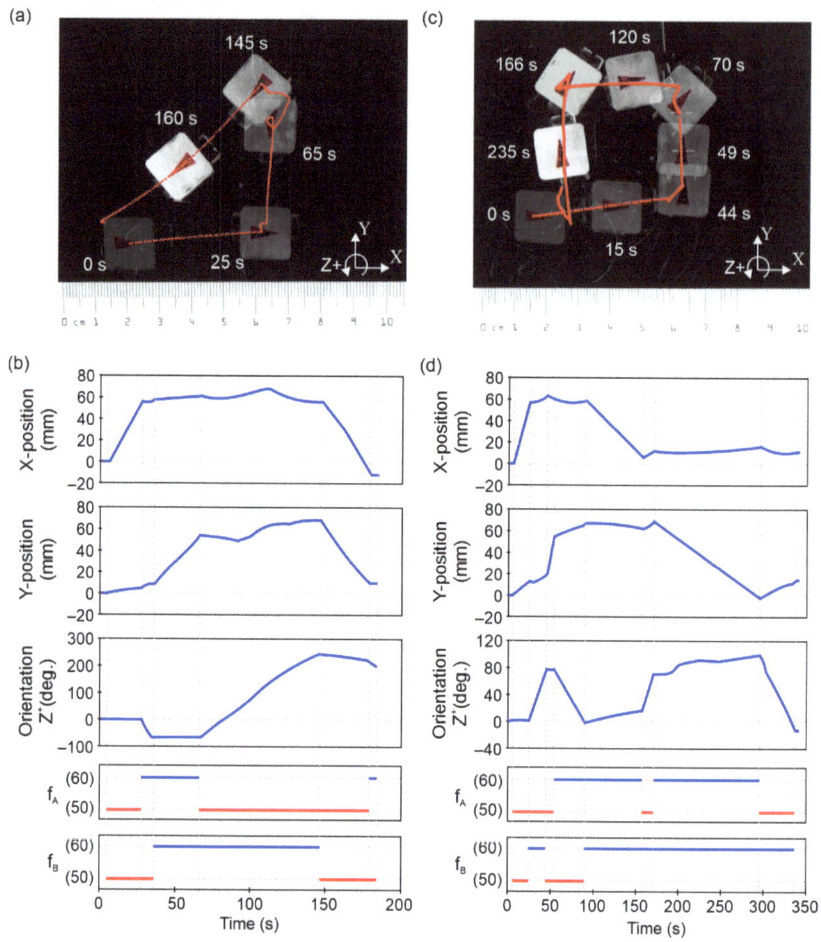

Figure 12. Results from the complex trajectory experiments, describing a 50 mm side right triangle (figures on the left) and a 40 mm side square (figures on the right). Top figures (**a**,**c**) show the robot position and orientation at some time instants of the preprogramed trajectory (red line). Bottom figures (**b**,**d**) detail the X- and Y-position and orientation of each trajectory vs. time and the sequence of resonant frequencies on plate A (f_A) and plate B (f_B) for actuation at the (50) or (60) modes.

As it can be seen in Figure 12a, the robot was able to describe a closed, right triangular trajectory of a 5 cm side in less than 200 s. The sequence of movements can be drawn from Figure 12b, where the position and orientation of the robot vs. time are depicted. The most challenging steps were a 330 deg. CCW turn (with less than 20 mm deviation in the X and Y axis) and a 65 mm long backward travel (with only 17 deg. orientation deviation). These deviations can be attributed to the mechanical stress of the wires.

Similar results were obtained with the square trajectory in Figure 12c. There, the robot performed both 90 deg. CW and CCW turns as well as 40 mm long forward and backward travels. As can be seen in Figure 12d, comparable deviations were measured in this trajectory, which demonstrated a promising trajectory control.

4.5. Performance Comparison

Table 5 summarizes a performance comparison of recent publications about miniaturized locomotion systems including our present robot. As can be seen, our proposed design is the only sub-gram robot with 2 degrees of freedom and a bidirectional motion among these references, featuring a minimum turning radius below 10 microns and a positional and rotational precision as low as 400 nm and 0.013 deg., respectively. In addition, the low-cost 3D printing fabrication technology used in our robot allows for an affordable platform in contrast to the more expensive MEMS or nanolithography printing technologies.

Table 5. Performance comparison of the most recent publications of miniaturized locomotion systems including our present robot. Unavailable data are indicated as "n.a.".

Reference	Size (mm^3)	Mass (mg)	Fabrication Technology	Actuation	DOF	Bidirectional	Autonomous	Tethered	Voltage (V)	Power (mW)	Payload (g)	Speed (BL/s)	Resolution (nm)
[8]	$2.5 \times 1.6 \times 0.7$	1	Nano 3DP	Magnetic	1	No	No	No	2	897	0	14.9	n.a.
[4]	$8.5 \times 4 \times 0.5$	10	MEMS	Electrostatic	1	No	Yes	No	50	0.01	0	7×10^{-4}	n.a.
[7]	$5 \times 5 \times 0.5$	18	MEMS	Electrostatic	1	No	No	Yes	80	0.08	0	0.2	n.a.
[6]	$4 \times 4 \times 5$	25	MEMS	Magnetic	1	No	No	No	n.a.	0.51	0	6.4	n.a.
[5]	$30 \times 15 \times 0.1$	64	Laminates	Piezoelectric	1	No	No	Yes	200	0.34	0.4	4	$2 \cdot 10^6$
[3]	$30 \times 15 \times 0.1$	65	Laminates	Piezoelectric	2	No	No	Yes	250	n.a.	0.18	7.5	n.a.
[27]	$15 \times 5 \times 0.5$	80	MEMS	Thermal	1	No	No	Yes	18	1100	2.5	0.4	n.a.
Present work	$20 \times 6 \times 0.85$	**200**	**3DP**	Piezoelectric	**2**	**Yes**	**No**	**Yes**	**10**	**10**	**25**	**0.1**	400
[11]	$12 \times 8 \times 6$	200	3DP	Piezoelectric	2	No	No	Yes	40	n.a.	13	5.2	n.a.
[3]	$24 \times 22 \times 0.1$	240	Laminates	Piezoelectric	2	No	Yes	No	250	397	1.66	1.13	n.a.
[15]	$20 \times 3 \times 1$	250	Glass + 3DP	Piezoelectric	1	Yes	No	Yes	30	n.a.	7.5	12.5	n.a.
[28]	$50 \times 50 \times 20$	1270	Laminates	Piezoelectric	2	Yes	Yes	No	200	50	1.35	10.0	1000
[18]	$50 \times 50 \times 20$	2400	Laminates	Piezoelectric	2	Yes	Yes	No	10	1660	0.44	0.6	n.a.
[28]	$58 \times 44 \times 12$	4.3×10^4	Machining	Piezoelectric	2	Yes	No	Yes	200	6666	200	8.9	440
[29]	$200 \times 200 \times 40$	5.6×10^6	Machining	Piezoelectric	2	Yes	No	Yes	200	n.a.	2.5×10^4	0.003	16

Our robot demonstrated the potential to become a fully autonomous motion system, thanks to a large carrying capacity in combination with a lower estimated power consumption. It is worth pointing out that two untethered and autonomous sub-gram robots have already been reported: the work by Hollar et al. [4] based on the MEMS technology featuring lower speeds and higher voltages, and the work by Liang et al. [3], which required excitation voltages as high as 250 V. Finally, our proposed design demonstrated submicrometric positional resolutions competing only with the heavier steel-based systems.

5. Conclusions

This work reports the design, fabrication, and characterization of a miniature 20 mm long sub-gram robot whose direction of motion can be controlled. A differential drive approach was implemented in a H-shaped 3D-printed motor platform by means of two-legged plate resonators, joined together at the middle length. The generation of the SWs on each plate was achieved by piezoelectric patches and the control of the motion path was attained by adjusting the parameters of the burst excitation signals such as the applied voltage, the number of applied cycles, or the driving frequency.

In terms of performance, bidirectional straight paths as long as 65 mm at 2 mm/s speed with only 10 V excitation were achieved, and furthermore, forward and backward precise steps as low as 1 µm. The turning of the robot could be controlled with turns as low as 0.013 deg. and an angular speed as high as 3 deg./s under the same conditions. With these optimizations, the proposed device demonstrated the execution of complex trajectories of more than 160 mm while carrying 70 times its own weight.

Regarding the lifetime of the system, mechanical fatigue fracture was not observed in our 3D printed platforms, likely due to the high stiffness of the glass-filled resin material. In addition, there are no moving parts in the design that can be exposed to a considerable wear except the leg tip. Although the robots were working for more than 100 h during the experiments without a noticeable impact on the leg tip, more tests should be conducted including different surfaces to determine the actual lifetime of the system.

Supplementary Materials: The following are available online at https://www.mdpi.com/article/10.3390/act10120335/s1, Video S1: Preprogrammed locomotion on a 50 mm side right triangular trajectory. Video S2: Preprogramed locomotion on a 40 mm side square trajectory.

Author Contributions: Conceptualization, J.L.S.-R.; Funding acquisition, J.H.-G. and J.L.S.-R.; Investigation, J.L.G.-C.; Project administration, J.H.-G. and J.L.S.-R.; Software, V.R.-D.; Supervision, J.H.-G. and J.L.S.-R.; Writing—original draft, V.R.-D.; Writing—review and editing, V.R.-D., J.H.-G. and J.L.S.-R. All authors have read and agreed to the published version of the manuscript.

Funding: This research was funded by regional JCCLM project SBPLY/17/180501/000139 and Grant RTI18-094960-B-100 by MCIN/AEI/10.13039/501100011033 and by FEDER "ERDF A way of making Europe".

Institutional Review Board Statement: Not applicable.

Informed Consent Statement: Not applicable.

Data Availability Statement: Not applicable.

Conflicts of Interest: The authors declare no conflict of interest.

References

1. Uchino, K. *MicroMechatronics*, 2nd ed.; CRC Press: Boca Raton, FL, USA, 2020; ISBN 978-0-367-20231-6.
2. St. Pierre, R.; Bergbreiter, S. Toward Autonomy in Sub-Gram Terrestrial Robots. *Annu. Rev. Control Robot. Auton. Syst.* **2019**, *2*, 231–252. [CrossRef]
3. Liang, J.; Wu, Y.; Yim, J.K.; Chen, H.; Miao, Z.; Liu, H.; Liu, Y.; Liu, Y.; Wang, D.; Qiu, W.; et al. Electrostatic Footpads Enable Agile Insect-Scale Soft Robots with Trajectory Control. *Sci. Robot.* **2021**, *6*, eabe7906. [CrossRef] [PubMed]
4. Hollar, S.; Flynn, A.; Bellew, C.; Pister, K.S.J. Solar Powered 10 Mg Silicon Robot. In Proceedings of the IEEE The Sixteenth Annual International Conference on Micro Electro Mechanical Systems, 2003. MEMS-03 Kyoto, Kyoto, Japan, 2–4 January 2003; pp. 706–711.
5. Wu, Y.; Yim, J.K.; Liang, J.; Shao, Z.; Qi, M.; Zhong, J.; Luo, Z.; Yan, X.; Zhang, M.; Wang, X.; et al. Insect-Scale Fast Moving and Ultrarobust Soft Robot. *Sci. Robot.* **2019**, *4*, eaax1594. [CrossRef] [PubMed]
6. Vogtmann, D.; Pierre, R.S.; Bergbreiter, S. A 25 MG Magnetically Actuated Microrobot Walking at Gt; 5 Body Lengths/Sec. In Proceedings of the 2017 IEEE 30th International Conference on Micro Electro Mechanical Systems (MEMS), Las Vegas, NV, USA, 22–26 January 2017; pp. 179–182.
7. Contreras, D.S.; Drew, D.S.; Pister, K.S.J. First Steps of a Millimeter-Scale Walking Silicon Robot. In Proceedings of the 2017 19th International Conference on Solid-State Sensors, Actuators and Microsystems (TRANSDUCERS), Kaohsiung, Taiwan, 18–22 June 2017; pp. 910–913.
8. St. Pierre, R.; Gosrich, W.; Bergbreiter, S. A 3D-Printed 1 mg legged microrobot running at 15 body lengths per second. In Proceedings of the 2018 Solid-State, Actuators, and Microsystems Workshop Technical Digest, Hilton Head, SC, USA, 21 May 2018; pp. 59–62.
9. Kim, D.; Hao, Z.; Mohazab, A.R.; Ansari, A. On the Forward and Backward Motion of Milli-Bristlebots. *Int. J. Non-Linear Mech.* **2020**, *127*, 103551. [CrossRef]
10. Li, J.; Pumera, M. 3D Printing of Functional Microrobots. *Chem. Soc. Rev.* **2021**, *50*, 2794–2838. [CrossRef] [PubMed]
11. Hao, Z.; Kim, D.; Mohazab, A.R.; Ansari, A. Maneuver at Micro Scale: Steering by Actuation Frequency Control in Micro Bristle Robots. In Proceedings of the 2020 IEEE International Conference on Robotics and Automation (ICRA), Paris, France, 31 May–31 August 2020; pp. 10299–10304.
12. Formlabs 3D Printing Materials Library. Available online: https://formlabs.com/uk/materials/ (accessed on 6 October 2021).
13. He, S.; Chen, W.; Tao, X.; Chen, Z. Standing Wave Bi-Directional Linearly Moving Ultrasonic Motor. *IEEE Trans. Ultrason. Ferroelectr. Freq. Control* **1998**, *45*, 1133–1139. [CrossRef] [PubMed]
14. Hariri, H.H.; Soh, G.S.; Foong, S.; Wood, K. Locomotion Study of a Standing Wave Driven Piezoelectric Miniature Robot for Bi-Directional Motion. *IEEE Trans. Robot.* **2017**, *33*, 742–747. [CrossRef]
15. Hernando-García, J.; García-Caraballo, J.L.; Ruiz-Díez, V.; Sánchez-Rojas, J.L. Comparative Study of Traveling and Standing Wave-Based Locomotion of Legged Bidirectional Miniature Piezoelectric Robots. *Micromachines* **2021**, *12*, 171. [CrossRef] [PubMed]
16. Dharmawan, A.G.; Hariri, H.H.; Foong, S.; Soh, G.S.; Wood, K.L. Steerable Miniature Legged Robot Driven by a Single Piezoelectric Bending Unimorph Actuator. In Proceedings of the 2017 IEEE International Conference on Robotics and Automation (ICRA), Singapore, 29 May–3 June 2017; pp. 6008–6013.
17. Baisch, A.T.; Ozcan, O.; Goldberg, B.; Ithier, D.; Wood, R.J. High Speed Locomotion for a Quadrupedal Microrobot. *Int. J. Robot. Res.* **2014**, *33*, 1063–1082. [CrossRef]
18. Hoover, A.M.; Steltz, E.; Fearing, R.S. RoACH: An Autonomous 2.4 g Crawling Hexapod Robot. In Proceedings of the 2008 IEEE/RSJ International Conference on Intelligent Robots and Systems, Nice, France, 7 September 2008; pp. 26–33.

19. Gogola, M.A.; Fischer, G.; Goldfarb, M.; Garcia, E. Development of Two Piezoelectrically Actuated Mesoscale Robot Quadrupeds. In Proceedings of the Microrobotics and Microassembly, SPIE, Boston, MA, USA, 21–22 September 1999; Volume 3834, pp. 76–84.
20. Rios, S.A.; Fleming, A.J.; Yong, Y.K. Miniature Resonant Ambulatory Robot. *IEEE Robot. Autom. Lett.* **2017**, *2*, 337–343. [CrossRef]
21. Ioi, K. A Mobile Micro-Robot Using Centrifugal Forces. In Proceedings of the 1999 IEEE/ASME International Conference on Advanced Intelligent Mechatronics (Cat. No.99TH8399), Atlanta, GA, USA, 19–23 September 1999; pp. 736–741.
22. Ruiz-Díez, V.; Hernando-García, J.; Toledo, J.; Ababneh, A.; Seidel, H.; Sánchez-Rojas, J.L. Piezoelectric MEMS Linear Motor for Nanopositioning Applications. *Actuators* **2021**, *10*, 36. [CrossRef]
23. Formlabs Rigid 10K Technical Data Sheet; Online. 2021. Available online: https://formlabs-media.formlabs.com/datasheets/2001479-TDS-ENUS-0.pdf (accessed on 30 November 2021).
24. Ruiz, V.; Hernando-García, J.; Ababneh, A.; Seidel, H.; Schmid, U.; Sánchez-Rojas, J.L. Modelling and Characterization of AlN-Actuated Microcantilevers Vibrating in the First in-Plane Mode. *Microsyst. Technol.* **2012**, *18*, 997–1001. [CrossRef]
25. Goldberg, B.; Zufferey, R.; Doshi, N.; Helbling, E.F.; Whittredge, G.; Kovac, M.; Wood, R.J. Power and Control Autonomy for High-Speed Locomotion with an Insect-Scale Legged Robot. *IEEE Robot. Autom. Lett.* **2018**, *3*, 987–993. [CrossRef]
26. Karpelson, M.; Wei, G.-Y.; Wood, R.J. Milligram-Scale High-Voltage Power Electronics for Piezoelectric Microrobots. In Proceedings of the 2009 IEEE International Conference on Robotics and Automation, Kobe, Japan, 12–17 May 2009; pp. 2217–2224.
27. Ebefors, T.; Mattsson, J.U.; Kälvesten, E.; Stemme, G. A Walking Silicon Micro-Robot. In Proceedings of the 10th International Conference on Solid-State Sensors and Actuators, Sendai, Japan, 1 January 1999; pp. 1202–1205.
28. Liu, Y.; Li, J.; Deng, J.; Zhang, S.; Chen, W.; Xie, H.; Zhao, J. Arthropod-Metamerism-Inspired Resonant Piezoelectric Millirobot. *Adv. Intell. Syst.* **2021**, *3*, 2100015. [CrossRef]
29. Deng, J.; Liu, Y.; Chen, W.; Yu, H. A XY Transporting and Nanopositioning Piezoelectric Robot Operated by Leg Rowing Mechanism. *IEEE/ASME Trans. Mechatron.* **2019**, *24*, 207–217. [CrossRef]

actuators MDPI

Article

Optimal Design of Magneto-Force-Thermal Parameters for Electromagnetic Actuators with Halbach Array

Zhihao Li [1], Qianqian Wu [1,*], Bilong Liu [1] and Zhaopei Gong [2]

1 School of Mechanical and Automotive Engineering, Qingdao University of Technology, Qingdao 266000, China; lizhihao914@163.com (Z.L.); liubilong@qut.edu.cn (B.L.)
2 Harbin Institute of Technology, School of Mechatronics Engineering, Harbin 150001, China; gongzp@hit.edu.cn
* Correspondence: wuqianqian@qut.edu.cn

Abstract: A magnetic levitation isolation system applied for the active control of micro-vibration in space requires actuators with high accuracy, linear thrust and low power consumption. The magneto-force-thermal characteristics of traditional electromagnetic actuators are not optimal, while actuators with a Halbach array can converge magnetic induction lines and enhance the unilateral magnetic field. To improve the control effect, an accurate magnetic field analytical model is required. In this paper, a magnetic field analytical model of a non-equal-size Halbach array was established based on the equivalent magnetic charge method and the field strength superposition principle. Comparisons were conducted between numerical simulations and analytical results of the proposed model. The relationship between the magnetic flux density at the air gap and the size parameters of the Halbach array was analyzed by means of a finite element calculation. The mirror image method was adopted to consider the influence of the ferromagnetic boundary on the magnetic flux density. Finally, a parametric model of the non-equal-size Halbach actuator was established, and the multi-objective optimization design was carried out using a genetic algorithm. The actuator with optimized parameters was manufactured and experiments were conducted to verify the proposed analytical model. The difference between the experimental results and the analytical results is only 5%, which verifies the correctness of the magnetic field analytical model of the non-equal-size Halbach actuator.

Keywords: electromagnetic actuators; Halbach arrays; magnetic levitation; magnetic field modeling; finite element simulation; analysis; multi-objective optimal design

Citation: Li, Z.; Wu, Q.; Liu, B.; Gong, Z. Optimal Design of Magneto-Force-Thermal Parameters for Electromagnetic Actuators with Halbach Array. *Actuators* **2021**, *10*, 231. https://doi.org/10.3390/act10090231

Academic Editor: Jose Luis Sanchez-Rojas

Received: 5 August 2021
Accepted: 6 September 2021
Published: 9 September 2021

1. Introduction

An electromagnetic actuator based on the Lorentz force principle is the core device of a magnetic levitation isolation platform, and it can eliminate the mechanical attachment between the stationary and moving parts of an actuator [1–3]. An actuator with high magnetic flux density, small volume and little hot loss is superior to others in obtaining suitable control performance. Compared with traditional actuators, the non-equal-size Halbach array can converge magnetic induction lines and enhance the unilateral magnetic field. Actuators with a Halbach array are characterized by small thrust fluctuation, high positioning accuracy, short response time, low influence of hysteresis effect and simple thrust control, all of which have considerable application prospects in relation to micro-vibration control in space [4–7].

Scholars have mainly focused on the dynamics model, the optimization of the magnetic circuit parameters, improving the electromagnetic thrust, reducing the thrust fluctuation, controlling the temperature rise effect, etc. [8]. Kou proposed a flat-type vertical-gap passive magnetic levitation vibration isolator (FVPMLVI) for an active vibration isolation system [9]. However, the author did not analyze the effect of the yoke portion on the magnetic field. Typical structure and rectangular structure voice coil motors were analyzed by their volumetric changes [10]. However, there was a lack of research into the dimensional parameters of permanent magnets. Lin designed a novel three-DOF spherical voice coil motor (VCM) and studied the simulation of the magnetic field distribution for rotation in relation to the x-axis and z-axis [11]. However, no specific analytical model of the magnetic field was involved. Wei developed an electromagnetic actuator that concurrently realized two working functions of vibration suppression and energy regeneration [12]. The experimental results also showed that the amplitude of the controlled object can be suppressed within 5.5%. However, the control accuracy of electromagnetic actuators has not been verified. Liu theorized a differential electromagnetic actuator for a magnetic levitation vibration isolation system. Linear output force and absolute velocity feedback were used to improve the vibration isolation performance [13]. However, a mathematical model of the magnetic field of the actuator has not been analyzed. A combination of the finite element method and the analytical method was discussed by Han, as was the eddy current loss of coil windings. The accuracy of thermal analysis predictions was verified through experiments [14]. A short-stroke planar motor with a non-integrated winding structure and an integrated winding structure was analyzed by Zhang [15]. The magnetic field distribution of the air gap was obtained using a finite element simulation. However, an analytical model is yet to be established. Considering the nonlinear saturation effect of ferromagnetic materials, a modified magnetic equivalent circuit (MEC) model was used by Liu to more accurately estimate the magnetic flux density at the air gap [16]. The magnetic leakage phenomenon in the middle of the magnetic conjugate was studied by Smith. The method of dividing permanent magnets and adjusting the pole distance was proposed to reduce magnetic leakage and thrust fluctuation [17]. A stacked giant magnetostrictive actuator was designed by He, and a magnetic field distribution model was established to accurately describe the magnetic field of the actuator, which improved the operating accuracy of the actuator in question [18]. A novel voice coil motor actuator that could generate a high actuating force was proposed by Kim; the magnetic field at the air gap was analyzed based on the charge model, which was verified using a 3D finite element method (FEM) simulation [19].

In order to improve the control effects, an accurate mathematical model of an actuator should be established by considering the impacts of the end effect and ferromagnetic boundary. In this paper, a non-equal-size actuator based on the Halbach array was designed for a magnetic levitation isolation platform. Compared with a traditional magnet, the Halbach array can converge magnetic induction lines and enhance the unilateral magnetic field. The magnetic flux density fluctuation of an equal-size Halbach array is higher than a non-equal-size Halbach array. By designing non-equal-size permanent magnets and optimizing the dimensional parameters, a uniformly distributed air-gap magnetic field can be obtained, which is conducive to improving the accuracy of an actuator and reducing its harmonic distortion. A magnetic field analytical model of a non-equal-size Halbach actuator was established and the magnetic flux density at the air gap was calculated. The analytical expression and numerical simulation of the magnetic flux density and fluctuation degree were compared between the Halbach array and the traditional permanent magnet. The influence of size parameters and the ferromagnetic boundary on the magnetic field of the Halbach array were studied. To obtain an actuator with uniform magnetic flux density distribution, high control accuracy, low power consumption and light mass, the multi-objective optimization design of a Halbach actuator was carried out and optimal magnetic-force-thermal parameters were obtained.

2. Modeling of the Magnetic Field of a Non-Equal-Size Halbach Array

2.1. Spatial Magnetic Field Analytical Model of Rectangular Permanent Magnets

The electric field distribution of a charged object in space can be considered as a superposition of an electric field intensity generated by electric charges. The magnetic field distribution of a uniformly magnetized rectangular permanent magnet can similarly be equated to a vector superposition of magnetic fields generated by a body magnetic charge inside the permanent magnet and surface magnetic charge on the surface [20]. The expressions for the body and surface charges can be written as follows:

$$\rho_m = -\mu_0 \nabla \cdot M \tag{1}$$

$$\sigma_m = M \cdot n = \frac{B_r}{\mu_0} \cdot n \tag{2}$$

where M means the magnetization intensity, n means the normal vector of the magnetic charge plane, B_r means the residual flux density of permanent magnets, and μ_0 means the vacuum permeability.

The rectangular permanent magnet is assumed to be uniformly magnetized. Therefore the volume magnetic charge density is 0, and the surface magnetic charge density is determined by the spatial magnetic field distribution of the rectangular permanent magnet.

There is no free current in the region of the static magnetic field generated by the rectangular permanent magnet; therefore, the expression of Maxwell Equations can be simplified as follows:

$$\nabla \times H = 0 \tag{3}$$

$$\nabla \cdot B = 0 \tag{4}$$

where B is the magnetic flux density and H is the magnetic field intensity.

The expressions for the scalar magnetic potential and magnetic properties of the magnetic field strength can be written as follows:

$$H = -\nabla \varphi_m \tag{5}$$

$$B = \mu_0 (H + M) \tag{6}$$

According to Equations (3)–(6), Equation (7) can be deduced as follows:

$$\nabla^2 \varphi_m = \nabla \cdot M \tag{7}$$

The integral expression for the scalar magnetic potential can be obtained according to Green's function.

$$\varphi_m(e_0) = \int G(e, e_0) \nabla' \cdot M(e_0) dv = \frac{1}{4\pi} \int_s \frac{M(e) \cdot n}{|e - e_0|} ds = \frac{1}{4\pi} \int_s \frac{\sigma_m}{|e - e_0|} ds \tag{8}$$

where e_0 means the observation point location vector and e means the field source point position vector.

According to Equation (8), the expression for the magnetic flux density of a rectangular permanent magnet in space can be written as follows:

$$B(e) = \frac{\mu_0}{4\pi} \int_s \frac{\sigma_m (e - e_0)}{|e - e_0|} ds \tag{9}$$

The equivalent magnetic charge models of the vertically magnetized and horizontally magnetized rectangular permanent magnets are shown in Figure 1. The vertically magnetized permanent magnets are equivalent to the upper and lower magnetic charge planes, and the horizontally magnetized permanent magnets are equivalent to the left and right magnetic charge planes [21].

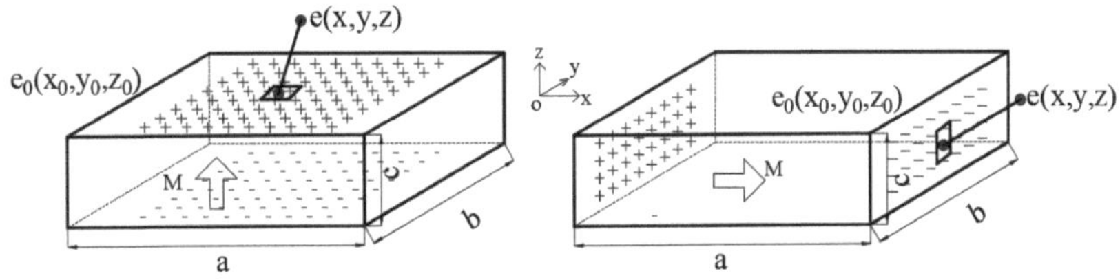

Figure 1. Equivalent magnetic charge model.

The geometric center of the rectangular permanent magnets is used as the origin of the coordinate axes. The magnetic flux density of two kinds of rectangular permanent magnets in space can be expressed as follows:

$$B_v(e) = \frac{\mu_0 \sigma_m}{4\pi} \sum_{k=1}^{2} (-1)^k \int_{-\frac{a}{2}}^{\frac{a}{2}} \int_{-\frac{b}{2}}^{\frac{b}{2}} \frac{e - e_0}{|e - e_0|^3} dx' y' \tag{10}$$

$$B_h(e) = \frac{\mu_0 \sigma_m}{4\pi} \sum_{i=1}^{2} (-1)^i \int_{-\frac{c}{2}}^{\frac{c}{2}} \int_{-\frac{b}{2}}^{\frac{b}{2}} \frac{e - e_0}{|e - e_0|^3} dx' z' \tag{11}$$

The magnetic field required for the Lorentz force is mainly provided by the *z*-axis component of the magnetic flux density at the air gap. The magnetic flux density of the *z*-axis component of the rectangular permanent magnets in two magnetizing directions is given as follows:

$$B_{z,v}(x,y,z) = \frac{\mu_0 \sigma_m}{4\pi} \sum_{i=1}^{2} \sum_{j=1}^{2} \sum_{k=1}^{2} (-1)^{i+j+k} \tan^{-1} \left[\frac{(x - x_i)(y - y_j)}{(z - z_k)} \frac{1}{[(x - x_i)^2 + (y - y_j)^2 + (z - z_k)^2]^{\frac{1}{2}}} \right] \tag{12}$$

$$B_{z,h}(x,y,z) = \frac{\mu_0 \sigma_m}{4\pi} \sum_{i=1}^{2} \sum_{j=1}^{2} (-1)^{i+k} \ln \frac{(y - y_1) + [(x - x_i)^2 + (y - y_1)^2 + (z - z_k)^2]^{\frac{1}{2}}}{(y - y_2) + [(x - x_i)^2 + (y - y_2)^2 + (z - z_k)^2]^{\frac{1}{2}}} \tag{13}$$

2.2. Parameterized Magnetic Field Model

Based on the equivalent magnetic charge method, the magnetic flux density of the vertically and horizontally magnetized rectangular permanent magnets at any point in space was deduced. In order to solve the magnetic flux density at the air gap of a non-equal-size Halbach array, a spatial coordinate system was established at the center of the *z*-axis of the air gap. The length, width and height of the vertical and horizontal magnetized permanent magnets are set as variables. A parameterized magnetic field model of the non-equal-size Halbach array was established, as shown in Figure 2.

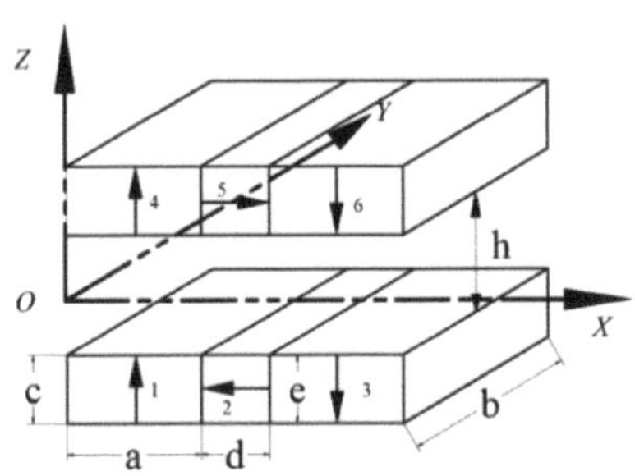

Figure 2. Parameterized magnetic field model of non-equal-size Halbach array.

The permanent magnets in different magnetizing directions in the magnetic field of Halbach array are numbered. According to the parameterized model, the boundary coordinates are determined as follows:

$$1\begin{cases} x_1 = 0; x_2 = a \\ y_1 = 0; y_2 = b \\ z_1 = -\frac{h}{2}; z_2 = -\frac{h}{2} - c \end{cases} \quad 2\begin{cases} x_1 = a; x_2 = a + d \\ y_1 = 0; y_2 = b \\ z_1 = -\frac{h}{2}; z_2 = -\frac{h}{2} - c \end{cases} \quad 3\begin{cases} x_1 = a + d; x_2 = 2a + d \\ y_1 = 0; y_2 = b \\ z_1 = -\frac{h}{2}; z_2 = -\frac{h}{2} - c \end{cases}$$

$$4\begin{cases} x_1 = 0; x_2 = a \\ y_1 = 0; y_2 = b \\ z_1 = \frac{h}{2}; z_2 = \frac{h}{2} + c \end{cases} \quad 5\begin{cases} x_1 = 0; x_2 = a \\ y_1 = 0; y_2 = b \\ z_1 = \frac{h}{2}; z_2 = \frac{h}{2} + c \end{cases} \quad 6\begin{cases} x_1 = 0; x_2 = a \\ y_1 = 0; y_2 = b \\ z_1 = \frac{h}{2}; z_2 = \frac{h}{2} + c \end{cases}$$

According to the superposition principle, the z-axis component of the flux density at the air gap can be expressed as follows:

$$B_z = \sum\nolimits_{m=1,3,4,6} B_{zm} + \sum\nolimits_{n=2,5} B_{zn} \quad (14)$$

Ansys was used to calculate the magnetic flux density at the air gap along the horizontal direction (x-axis) and the vertical direction (z-axis). A comparison between the analytical results and the simulation results are shown in Figure 3.

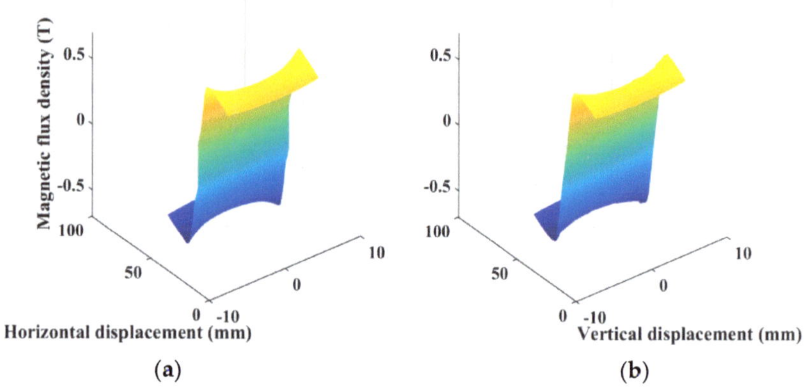

Figure 3. (**a**) Analytical results; (**b**) Simulation results.

The difference between the analytical value and the finite element simulation value of the magnetic flux density along the vertical direction is about 5%. The deviation between the analytical value and the simulation value along the horizontal direction is about 3%. Due to the end effect of the permanent magnet, the deviation between the two calculated results is 5–10%, which is within the allowable error range. It can be concluded that the magnetic field analytical model of the parameterized Halbach actuator is accurate.

2.3. Comparison of Magnetic Flux Density between the Non-Equal-Size Halbach Array, Equal-Size Halbach Array and Conventional Magnet Groups

The electromagnetic actuator is required to enhance the intensity of the magnetic field and reduce the thrust fluctuation as much as possible, which requires a high magnetic flux density at the air gap and a uniform distribution of magnetic induction lines. The non-equal-size Halbach array, equal-size Halbach array and conventional array models are displayed in Figure 4. The analytical results and the simulation results of the magnetic field at the air gap are shown in Figure 5.

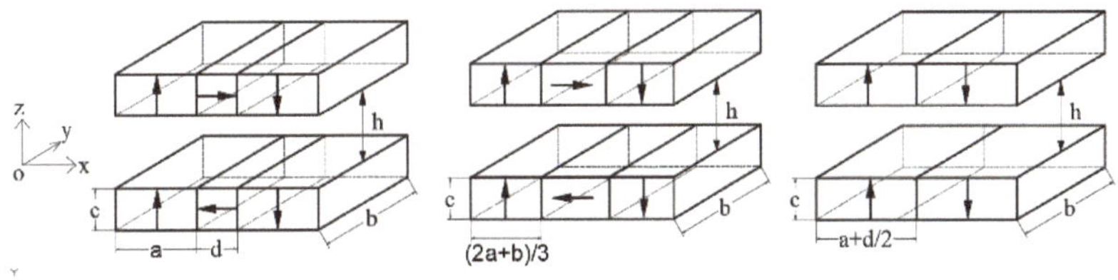

Figure 4. Different array models.

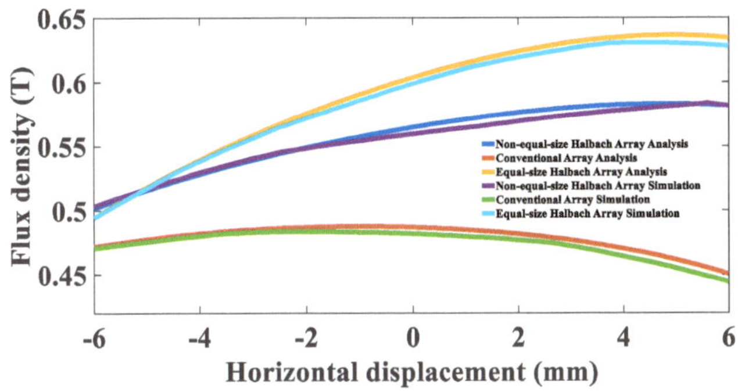

Figure 5. The magnetic flux density of the non-equal-size Halbach array, equal-size Halbach array and a conventional array.

The average magnetic flux density of the equal-size Halbach array and non-equal-size Halbach array is about 23 and 17% higher than conventional array, respectively. It indicates that the actuator of the Halbach array outputs more thrust under the same current excitation.

The expression of magnetic flux density fluctuation is calculated as follows:

$$\Delta B = \left| \frac{B_{max} - B_{min}}{B_{max}} \right| \times 100\% \tag{15}$$

where B_{max} means the air gap flux density maximum and B_{min} means the air gap flux density minimum.

Within the effective stroke of the coil, the fluctuation of the non-equal-size Halbach array and the equal-size Halbach array are shown in Table 1.

Table 1. Magnetic flux density fluctuation.

	Non-Equal-Size	Equal-Size
Simulation result	13.67%	21.57%
Analysis result	13.94%	22.35%

It can be seen that the magnetic flux intensity of the equal-size Halbach array is a little greater than the non-equal-size Halbach array. However, the end effect and the fluctuation of the magnetic flux density of the non-equal-size Halbach array can be reduced more significantly.

3. Effect of the Dimensional Parameters of the Halbach Array

3.1. Effect of the Width of the Horizontal Magnetized Permanent Magnet

The purpose of horizontal magnetized permanent magnets is to enhance the magnetic flux density and reduce the effect of the permanent magnet end effect at the air gap. The sizes of the vertical magnetized permanent magnet and the air gap distance remain unchanged. Then, the relationship between size d of the horizontal magnetized permanent magnet and the magnetic flux density at the air gap is apparent (Figure 6).

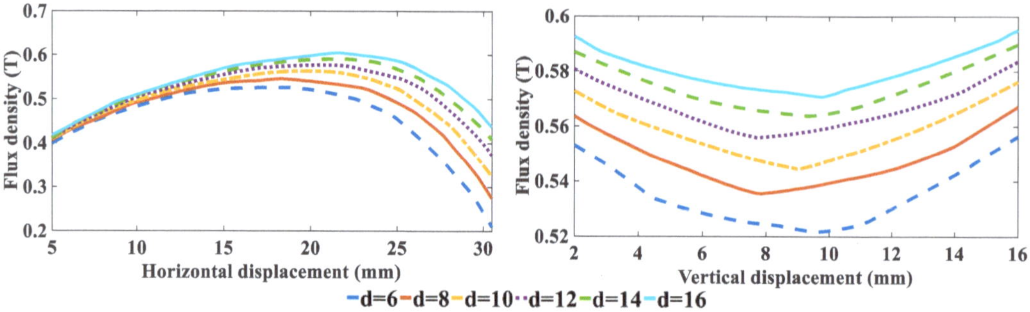

(a) (b)

Figure 6. The relationship between size d of horizontal magnetized permanent magnet and the magnetic flux density at the air gap: (**a**) Horizontal displacement; (**b**) Vertical displacement.

3.2. Effect of Width Ratio of the Vertically and Horizontally Magnetized Permanent Magnets

The maximum width of the actuator is set as the same width as the Halbach array. Additionally, the total width, length and height of the Halbach array are constant. The relation between the width ratio and the magnetic flux density was then simulated, as shown in Figure 7.

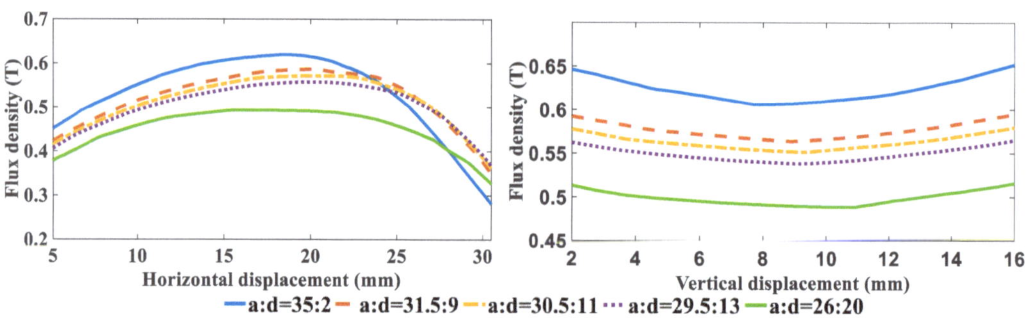

(a) (b)

Figure 7. The relation between width ratio and magnetic flux density: (**a**) Horizontal displacement; (**b**) Vertical displacement.

In the area far from the horizontal magnetized permanent magnet, the vertical magnetized permanent magnet dominates the magnetic field. Increasing the width ratio enhances the magnetic field along the horizontal direction. In the area near the horizontal magnetized permanent magnet, the increase in the magnetic flux density caused by the increase in the width of the horizontal magnetized permanent magnet is higher than the decay caused by the decrease in the width of the vertical magnetized permanent magnet. Thus, the magnetic flux density in the horizontal direction decreases with the increase in the width ratio. The standard variance of magnetic flux density in the x-axis and z-axis directions is shown in Figure 8.

When the text states a:d = 3:1, the magnetic flux density along the x-axis direction is stronger and the magnetic induction lines are more uniform. When the text states a:d = 2.5–3.5, the magnetic flux density along the z-axis direction has a smaller standard variance, which proves that the magnetic induction lines are more uniformly distributed.

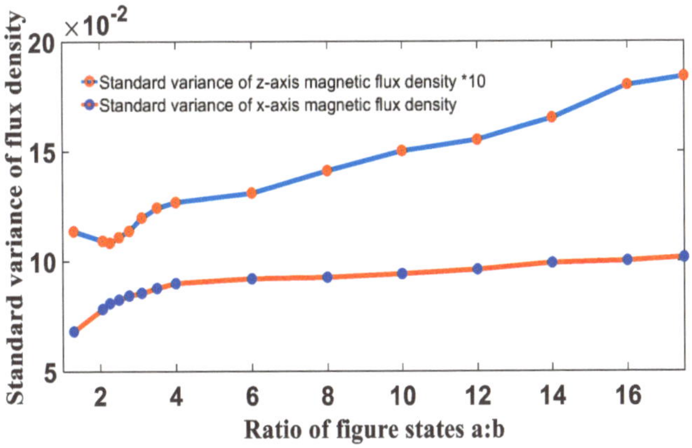

Figure 8. Standard variance of magnetic flux density.

4. Effect of Ferromagnetic Boundary

The designed actuator has bilateral ferromagnetic boundaries, as shown in Figure 9. It contains six rectangular permanent magnets and two magnetic yokes at the top and bottom. The existence of ferromagnetic boundaries has a certain influence on the magnetic flux density at the air gap compared to that without ferromagnetic boundaries [22]. The influence of the ferromagnetic boundary on the original magnetic field was calculated using the mirror method, which used the equivalent magnetic charge behind the boundary to equate its influence. The actual spatial distribution of the magnetic field is formed by the superposition of the original magnetic field and the mirror field.

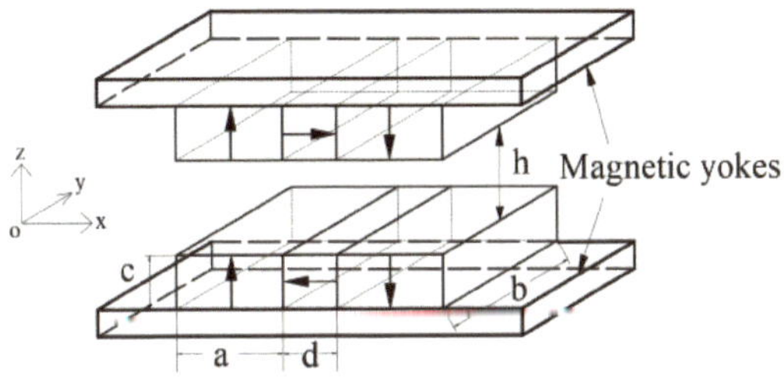

Figure 9. Halbach array with bilateral ferromagnetic boundaries.

A schematic diagram of the mirror image of the magnetic charge surface of the Halbach array under bilateral ferromagnetic boundary conditions is shown in Figure 10. H is the distance between the bilateral ferromagnetic boundaries. Theoretically, there are infinite sets of mirror images of the magnetic charge surface.

The analytical expression for the magnetic flux density at the air gap can be calculated according to the mirror image principle and the superposition principle as follows:

$$B(x,y,z) = \sum_{m=1,3}\sum_{i=0}^{\infty} B_{m,vertical}(x,y,z \pm iH) + \sum_{i=0}^{\infty} B_{horizontal}(x,y,z \pm iH) \tag{16}$$

The ferromagnetic boundary material parameters are shown in Table 2.

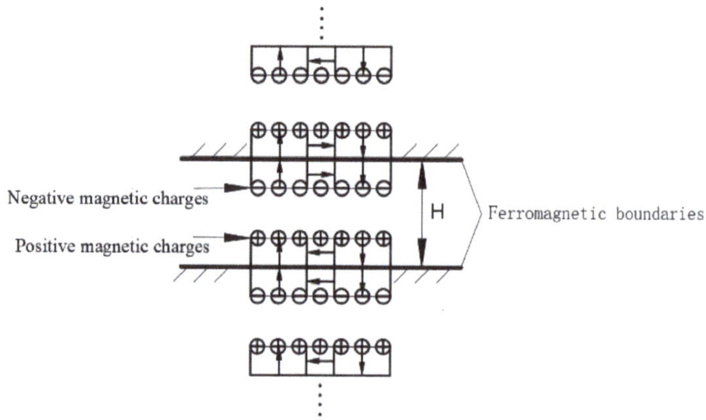

Figure 10. Schematic diagram of the mirror image of the magnetic charge surface.

Table 2. Ferromagnetic material parameters.

Parameters	Numerical Value	Unit
a	30	mm
b	60	mm
c	20	mm
d	11	mm
e	20	mm
h	16	mm
B_r	1.1	T
μ_0	$4\pi \times 10^{-7}$	H/m

A comparison between the simulation results and the analytical results of the magnetic flux density at the air gap magnetic field is shown in Figure 11. It can be seen that the average magnetic flux density along the *z*-axis with a ferromagnetic boundary is 22% higher than that without a ferromagnetic boundary. This is because the ferromagnetic boundary can concentrate magnetic force lines and enhance magnetic field strength. The maximum deviation between the simulation and the analytical calculation is 2.2%. Additionally, the analytical result derived using the mirror method is in good agreement with the finite element simulation value. Since one set of mirror images is superimposed for analytical value, the analytical value is smaller than the simulation value.

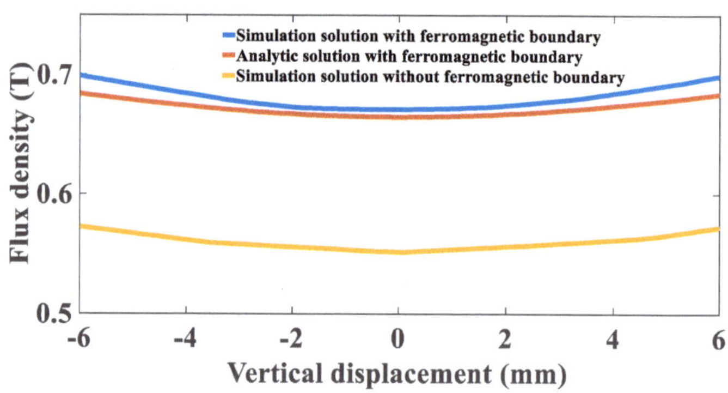

Figure 11. Comparison between magnetic flux density along *z*-axis with and without ferromagnetic boundary.

5. Optimized Design

The mechanical structure of the actuator consists of a kinematic coil and a stator Halbach array, and the parametric model is shown in Figure 12.

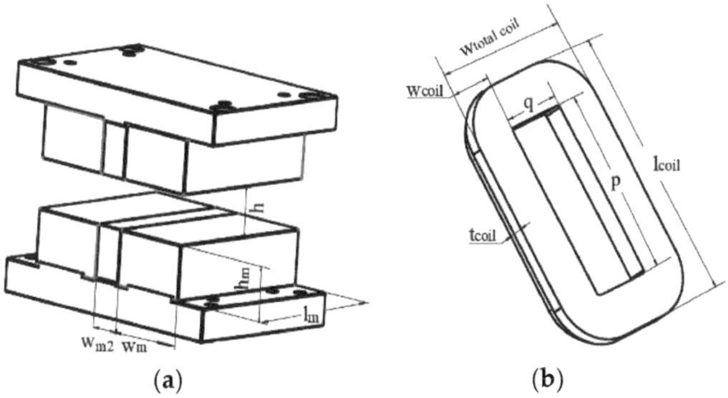

Figure 12. Parametric model of the actuator: (**a**) Stator Halbach array; (**b**) Kinematic coil.

Two objectives are needed for optimization. First, the magnetic flux density at the air gap is maximized and uniformly distributed. Second, the minimum coil mass and minimum thermal power consumption should be obtained. A genetic algorithm was adopted to obtain the actuator with optimal magnetic-force-thermal parameters.

According to the magnetic field analytical model of the non-equal-size Halbach array, and when considering the influence of the ferromagnetic boundary, the approximate relationship between the magnetic field at the center of the air gap and the parameters of the actuator is as follows:

$$\begin{gathered} B_Z(w_m, w_{m2}, l_m, h_m, h) = \frac{2B_r}{\pi}\tan^{-1}\left[\frac{w_m l_m}{s}\frac{1}{[w_m^2+l_m^2+h^2]^{1/2}}\right] \\ +\sum_{i=1}^{\infty}\frac{2B_r}{B_r}\tan^{-1}\left[\frac{w_m l_m}{h\pm 2i(h+2h_m)}\right]\left[\frac{1}{[w_m^2+l_m^2+(h\pm 2i(h+2h_m))^2]^{\frac{1}{2}}}\right] \end{gathered} \tag{17}$$

where w_m means the width of the vertically magnetized permanent magnet, w_{m2} means the width of the horizontally magnetized permanent magnet, l_m means the length of the vertically magnetized permanent magnet, h_m means the height of the vertically magnetized permanent magnet and h means the length of the air gap.

When the width of the horizontal magnetized permanent magnet is greater than 10mm, and the width ratio of vertical and horizontal magnetized permanent magnet is 2.5–3.5, the magnetic field is smooth and uniform, and magnetic flux density can reach its maximum value.

The optimization objectives are expressed as follows:

$$Opt.\begin{cases} Q_{min,coil} = p\dfrac{16I^2\rho_r V_{coil}}{\pi^2 d_{coil}^4} \\ M_{min,coil} = (1-p)\rho\eta V_{coil} \end{cases} \tag{18}$$

where I is the coil current, ρ_r is the wire resistivity, V_{coil} is the coil volume, d_{coil} is the wire diameter, ρ is the wire density and η is the copper fill rate. The weight is defined as p and can be set as 0.6 for the actuator.

The dimension constraints are given as Equation (19), including the relative position conditions of the coil and the Halbach array and actuator boundary dimension constraints.

$$s.t.\begin{cases} h \geq t_{coil} + 2s + f \\ w_m \geq w_{coil} + 2s \\ l_{coil} \geq l_m + 2w_{coil} + 2s \\ w_{total\ coil} \geq 2w_{coil} + w_{m2} + 2s \\ x_1 \geq 2w_m + w_{m2} \\ x_1 \geq 2w_m + w_{m2} \\ x_3 \geq l_{coil} \end{cases} \quad (19)$$

where t_{coil} means the coil thickness, f means the coil winding case thickness, w_{coil} means the coil single side width, $w_{total\ coil}$ means the total coil width, l_{coil} means the coil length and x_1, x_2, x_3 means the maximum design size of the actuator.

Based on a genetic algorithm, the optimal air gap size, minimum coil mass and minimum thermal dissipation are calculated. A comparison with the parameters before and after optimization is shown in Table 3.

Table 3. Comparison of parameters before and after optimization.

	Actuator Parameters	Before Optimization	After Optimization	Unit
Halbach array	Total length	58	60	mm
	Total width	70	72	mm
	Height	18	20	mm
	Width of vertical permanent magnet	25	30.5	mm
	Width of horizontal permanent magnet	20	11	mm
	Air gap length	20	15	mm
Coil	Total length	105.6	99	mm
	Total width	46	56	mm
	Number of turns	88	63	turn
Actuator	Effective stroke	±3	±3	mm
	Force constants	5.33	6.98	N/A
	Continuous thrust	42.68	56.5	N
	Mass of the coil	380	323.04	g
	Consumption of the coil	17.9	16.32	W

The output force range of one single actuator is 0.1–50 N, and the current accuracy is 0.005A; force constants can be estimated as follows:

$$K = \frac{F}{I} = \frac{0.1}{0.005} = 20 > 6.98 \quad (20)$$

The force constant after optimization is smaller than the design index. According to the minimum current condition of the system, the control requirement of minimum Lorentz force can be satisfied.

6. Experimental Section

An actuator prototype was manufactured with optimized parameters, and the experimental setup is shown in Figure 13. The actuator was fixed on a high-precision displacement platform and a Gauss meter probe was mounted on a motion axis with a laser displacement sensor. The position of the Gauss meter probe at the air gap was adjusted by a real-time position control system to measure the magnetic flux density along the z-axis and x-axis directions at the air gap, respectively.

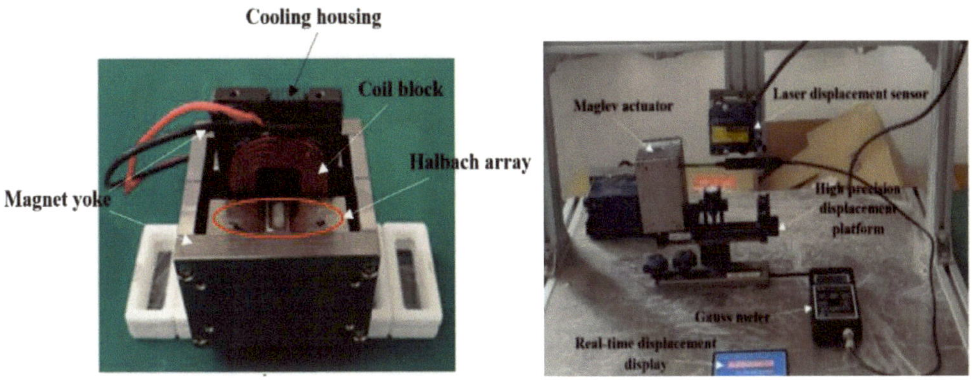

(a) (b)

Figure 13. Experimental verification: (**a**) Actuator prototype; (**b**) Experimental test platform.

The measured results were compared with the finite element simulation results and analysis results, as shown in Figure 14. Within the effective stroke of the actuator, the experimental results show that the average air gap flux density in the *z*-axis direction is 0.85T, with a flux density fluctuation of 4%, while the average flux density in the *x*-axis direction is 0.86T, with a flux density fluctuation of 3.8%.

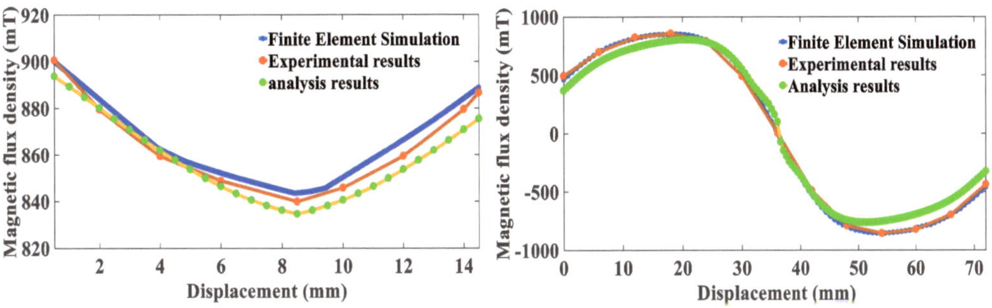

(a) (b)

Figure 14. Comparison between simulation, analysis and experimental results at the air gap: (**a**) *z*-axis; (**b**) *x*-axis.

7. Conclusions

A magnetic field analytical model of a non-equal-size Halbach array was established based on the equivalent magnetic charge method and the field superposition principle. The accuracy of the analytical model was verified by comparing it with Maxwell simulation results. We studied the influences of the size parameters of the Halbach array on the magnetic flux density at the air gap. A parametric model of the non-equal-size Halbach actuator was developed, and multi-objective optimization design was carried out to obtain optimal magneto-force-thermal parameters. The prototype of the actuator with optimized parameters was manufactured and experiments were conducted to verify the magnetic field analytical model. The force constant of the actuator is 6.98 N/A, which meets the minimum control force requirement. Within the effective stroke of the actuator, the average magnetic flux density at the air gap is 0.85T, which is 5% different from the analysis results. The simulation results are in high agreement with the analytical results. The experiment results show that the proposed magnetic field analytical model is accurate. A precision dynamic model of the suggested magnetic levitation vibration isolation system with a Halbach array will be studied by merging the proposed analytic magnetic field model in the next step.

Author Contributions: Conceptualization, Z.L. and Q.W.; methodology, Z.L.; software, Z.L.; validation, Z.L.; formal analysis, Z.L.; investigation, Z.G.; resources, B.L.; data curation, Z.L.; writing—original draft preparation, Z.L.; writing—review and editing, B.L and Q.W.; visualization, Z.L.; supervision, Z.G.; project administration, B.L.; funding acquisition, Q.W. All authors have read and agreed to the published version of the manuscript.

Funding: This research was funded by National Natural Science Foundation of China, grant number 51905288, Applied Basic Research Program of Science and Technology of Qingdao, grant number 19-6-2-61-cg and Open Project of Key Laboratory of Industrial Fluid Energy Conservation and Pollution Control, Ministry of Education.

Institutional Review Board Statement: Not applicable.

Informed Consent Statement: Not applicable.

Data Availability Statement: The data presented in this study are available on request from the corresponding author.

Conflicts of Interest: The authors declare no conflict of interest.

References

1. Poletkin, K.V.; Asadollahbaik, A.; Kampmann, R.; Korvink, J. Levitating Micro-Actuators: A Review. *Actuators* **2018**, *7*, 17. [CrossRef]
2. Wu, Q.; Cui, N.; Zhao, S.; Zhang, H.; Liu, B. Modeling and Control of a Six Degrees of Freedom Maglev Vibration Isolation System. *Sensors* **2019**, *19*, 3608. [CrossRef] [PubMed]
3. Jansen, J.H.; Van Lierop, C.N.; Lomonova, E.; Vandenput, A.J.A. Magnetically Levitated Planar Actuator with Moving Magnets. *IEEE Trans. Ind. Appl.* **2008**, *44*, 1108–1115. [CrossRef]
4. Kamesh, D.; Pandiyan, R.; Ghosal, A. Modeling, design and analysis of low frequency platform for attenuating micro-vibration in spacecraft. *J. Sound Vib.* **2010**, *329*, 3431–3450. [CrossRef]
5. Zhu, H.; Teo, T.J.; Pang, C.K. Design and Modeling of a Six-Degree-of-Freedom Magnetically Levitated Positioner Using Square Coils and 1-D Halbach Arrays. *IEEE Trans. Ind. Electron.* **2016**, *64*, 440–450. [CrossRef]
6. Meng, B.; Zhu, C.; Xu, H.; Dai, M.; Li, S. Analytical and Experimental Investigations of Novel Maglev Coupling Based on Opposed Halbach Array for a 2D Valve. *Actuators* **2021**, *10*, 61. [CrossRef]
7. Ham, C.; Ko, W.; Lin, K.-C.; Joo, Y. Study of a Hybrid Magnet Array for an Electrodynamic Maglev Control. *J. Magn.* **2013**, *18*, 370–374. [CrossRef]
8. Chai, J.W.; Gui, X. Review of voice coil motor structure optimization and application. *Trans. China Electrotech. Soc.* **2021**, *36*, 1113–1125. [CrossRef]
9. Kou, B.; Zhou, Y.; Yang, X.; Xing, F.; Zhang, H. Electromagnetic and Mechanical Characteristics Analysis of a Flat-Type Vertical-Gap Passive Magnetic Levitation Vibration Isolator. *Shock. Vib.* **2016**, *2016*, 1–12. [CrossRef]
10. Kim, J.-Y.; Ahn, D. Analysis of High Force Voice Coil Motors for Magnetic Levitation. *Actuators* **2020**, *9*, 133. [CrossRef]
11. Lin, Y.-H.; Liu, C.-S.; Yeh, C.-N. Design and Simulation of Novel 3-DOF Spherical Voice Coil Motor. *Actuators* **2021**, *10*, 155. [CrossRef]
12. Wei, W.; Li, Q.; Xu, F.; Zhang, X.; Jin, J.; Jin, J.; Sun, F. Research on an Electromagnetic Actuator for Vibration Suppression and Energy Regeneration. *Actuators* **2020**, *9*, 42. [CrossRef]
13. Liu, Y.; Tan, J.B.; Wang, L. Active control mechanism of vibration isolation foundation for very large optical instruments with differential electromagnetic actuators. *Opt. Precis. Eng.* **2007**, *15*, 1602–1608. [CrossRef]
14. Han, B.C.; Peng, S.; He, Z.; Liu, X.; Zhang, X. Calculation and thermal analysis of eddy current losses in the windings of magnetic levitation controlled torque gyro high-speed motors. *Opt. Precis. Eng.* **2020**, *28*, 130–140. [CrossRef]
15. Zhang, H.; Kou, B.Q.; Zhang, H.L.; Jin, Y.X.; Zhang, C.N.; Li, L.Y. Structural Optimization of an Integrated Winding Structure Short-Stroke Planar Motor Driven by Square DC Coils and Permanent Magnets. *Appl. Mech. Mater.* **2013**, *416–417*, 221–226. [CrossRef]
16. Liu, Y.; Zhang, M.; Zhu, Y.; Yang, J.; Chen, B. Optimization of Voice Coil Motor to Enhance Dynamic Response Based on an Improved Magnetic Equivalent Circuit Model. *IEEE Trans. Magn.* **2011**, *47*, 2247–2251. [CrossRef]
17. Smith, K.J.; Graham, D.; Neasham, J.A. Design and Optimization of a Voice Coil Motor with a Rotary Actuator for an Ultrasound Scanner. *IEEE Trans. Ind. Electron.* **2015**, *62*, 7073–7078. [CrossRef]
18. He, Z.B.; Rong, C.; Li, D.W.; Xue, G.M.; Zheng, J.W. Modeling and analysis of magnetic field distribution of stacked super magnetostrictive actuators. *Opt. Precis. Eng.* **2017**, *25*, 2347–2358. [CrossRef]

19. Kim, M.H.; Kim, H.Y.; Kim, H.C.; Ahn, D.; Gweon, D.-G. Design and Control of a 6-DOF Active Vibration Isolation System Using a Halbach Magnet Array. *IEEE/ASME Trans. Mechatron.* **2016**, *21*, 2185–2196. [CrossRef]
20. Gou, X.F.; Yang, Y.; Zheng, X.J. Analytic expression of magnetic field distribution of rectangular permanent magnets. *Appl. Math. Mech.* **2004**, *25*, 297–306. [CrossRef]
21. Zhang, H. Basic Research on Multi-Degree-of-Freedom Magnetic Levitation Micro-Motion Table. Ph.D. Dissertation, Harbin Institute of Technology, Harbin, China, 2014.
22. Kou, B.; Zhang, H.; Li, L. Analysis and Design of a Novel 3-DOF Lorentz-Force-Driven DC Planar Motor. *IEEE Trans. Magn.* **2011**, *47*, 2118–2126. [CrossRef]

actuators MDPI

Article

Using the Nonlinear Duffing Effect of Piezoelectric Micro-Oscillators for Wide-Range Pressure Sensing

Tobias Zengerle [1,*], Michael Stopp [1], Abdallah Ababneh [1,2] and Helmut Seidel [1]

1 Chair of Micromechanics, Microfluidics and Microactuators, Campus A5.1, Saarland University, 66123 Saarbrücken, Germany; michael.stopp@lmm.uni-saarland.de (M.S.); a.ababneh@yu.edu.jo (A.A.); seidel@lmm.uni-saarland.de (H.S.)
2 Electronic Engineering Departement, Hijjawi Faculty for Engineering Tech., Yarmouk University, Irbid 21163, Jordan
* Correspondence: t.zengerle@lmm.uni-saarland.de

Abstract: This paper investigates the resonant behaviour of silicon-based micro-oscillators with a length of 3600 μm, a width of 1800 μm and a thickness of 10 μm over a wide range of ambient gas (N_2) pressures, extending over six orders of magnitude from 10^{-3} mbar to 900 mbar. The oscillators are actuated piezoelectrically by a thin-film aluminium-nitride (AlN) layer, with the cantilever coverage area being varied from 33% up to 100%. The central focus is on nonlinear Duffing effects, occurring at higher oscillation amplitudes. A theoretical background is provided. All relevant parameters describing a Duffing oscillator, such as stiffness parameters for each coverage size as well as for different bending modes and more complex modes, are extracted from the experimental data. The so-called 2nd roof-tile-shaped mode showed the highest stiffness value of $-97.3 \cdot 10^7$ m^{-2}s^{-2}. Thus, it was chosen as being optimal for extended range pressure measurements. Interestingly, both a spring softening effect and a spring hardening effect were observed in this mode, depending on the percentage of the AlN coverage area. The Duffing-effect-induced frequency shift was found to be optimal for obtaining the highest pressure sensitivity, while the size of the hysteresis loop is also a very useful parameter because of the possibility of eliminating the temperature influences and long-term drift effects of the resonance frequency. An reasonable application-specific compromise between the sensitivity and the measurement range can be selected by adjusting the excitation voltage, offering much flexibility. This novel approach turns out to be very promising for compact, cost-effective, wide-range pressure measurements in the vacuum range.

Keywords: MEMS oscillator; nonlinearity; pressure sensor; roof-tile-shaped mode; AlN

Citation: Zengerle, T.; Stopp, M.; Ababneh, A.; Seidel, H. Using the Nonlinear Duffing Effect of Piezoelectric Micro-Oscillators for Wide-Range Pressure Sensing. *Actuators* **2021**, *10*, 172. https://doi.org/10.3390/act10080172

Academic Editor: Micky Rakotondrabe

Received: 14 June 2021
Accepted: 22 July 2021
Published: 24 July 2021

1. Introduction

Micromechanical oscillators enjoy a wide range of applications in many areas. In consumer and automotive electronics, gyroscopes and accelerometers have become omnipresent [1]. Applications that are more specific can be found as microbalances in bio analytics or as sensing elements in scanning probe microscopy (SPM) [2]. In the upcoming field of energy autonomous systems, piezoelectric oscillators are used as vibrational energy harvesters [3]. In some cases and under certain conditions, micro-oscillators may exhibit nonlinear behaviour. Such nonlinearities have been studied and described, e.g., in signal processing and amplification [4,5], or else in pressure sensing [6–8] and gas sensing applications [9], among others. In this context, nonlinearities can strongly affect the frequency response of micro-oscillators regarding their dynamic range and damping behaviour under varying ambient pressure and can therefore significantly increase the sensitivity of oscillator devices [3,4,10]. MEMS oscillators are, e.g., used for leakage detection [11] in autonomous systems which can be accessed via wireless networks [12] (e.g., tire pressure sensors).

This work focusses on the investigation of the nonlinear Duffing effect of micromechanical silicon cantilevers driven by piezoelectric thin film actuators. In order to understand the theoretical background, the description of electrostatically driven Duffing oscillators known from the literature [13–15] is adapted and expanded to oscillators with piezoelectric actuation. From this theory, the frequency response is derived and the coefficients for linear and nonlinear damping can be determined. The theoretical results are compared with experimentally obtained values for one type of oscillator, where the piezoelectric thin film actuation area is varied, covering 33%, 50%, or 100% of the cantilever. In contrast to previous works on the pressure dependence of nonlinear micro-oscillators [6,7], we use the frequency shift and the frequency hysteresis by the duffing nonlinearity as correlating quantity to the ambient pressure. Therefore, the sensor response is much higher compared to the classical measurement principle based on the frequency shift of linear damped oscillators.

2. Theory

2.1. Nonlinear Duffing Oscillator

The model for a linear damped harmonic oscillator driven by a periodic force $F(t)$ with actuation amplitude F_{piezo} and angular frequency ω (=$2\pi f$) is given by the following differential equation:

$$\ddot{A} + \lambda\dot{A} + {\omega_0}^2 A = \frac{F(t)}{m} = \frac{F_{\text{piezo}}}{m} \cdot \sin(\omega t) \; with \; {\omega_0}^2 = \frac{k}{m}, \tag{1}$$

where $A(t)$, λ, ω_0, $F(t)$, m, k are the oscillation amplitude of the movement, the linear damping coefficient, the mechanical angular resonant frequency, the actuation force, the mass and the linear stiffness parameter of the oscillator, respectively.

Nonlinear behaviour in its simplest form is included by adding a cubic term for the displacement A together with the nonlinear stiffness parameter β, which leads to an additional restoring force (see Figure 1). This results in a displacement depending on the spring constant, leading to a tilting response curve. Thus, we obtain the well-known Duffing equation [14,16]:

$$\ddot{A} + \lambda\dot{A} + {\omega_0}^2 A + \beta A^3 = \frac{F_{\text{piezo}}}{m} \cdot \sin(\omega t) \tag{2}$$

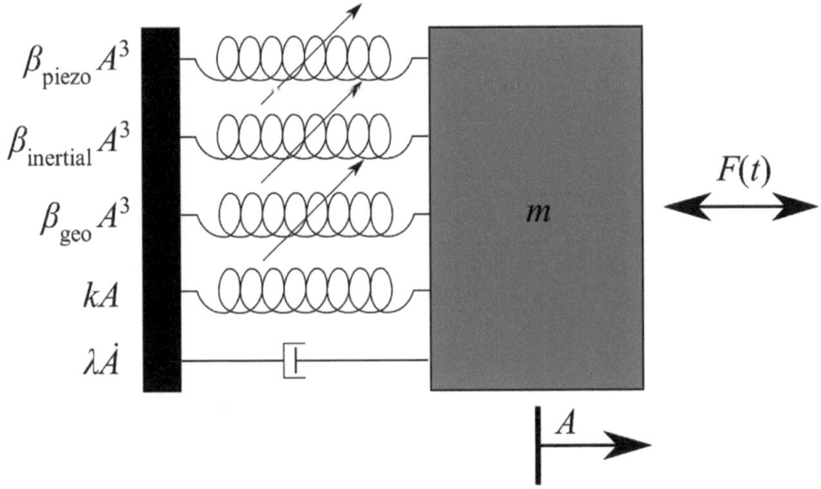

Figure 1. Schematic picture of a piezoelectrically actuated micro-oscillator with force $F(t)$ and mass m. The moving mass is attached to a fixed anchorage via a linear damper and four springs representing the linear and the three nonlinear restoring forces.

This nonlinear differential equation can be solved by the homotopy analysis method (HAM) to derive the frequency response of the oscillating amplitude A [17]:

$$A^2 = \frac{F_{\text{piezo}}{}^2}{m^2\left[\left(\omega^2 - k - \frac{3}{4}\beta A^2\right)^2 + (\lambda\omega)^2\right]} \tag{3}$$

Depending on the sign of the nonlinear stiffness parameter β, the resonant curve tilts to the left ($\beta < 0$) or to the right ($\beta > 0$) (see Figure 2a). In terms of a mechanical oscillating system, a spring softening ($\beta < 0$) or else a spring hardening ($\beta > 0$) effect can be observed.

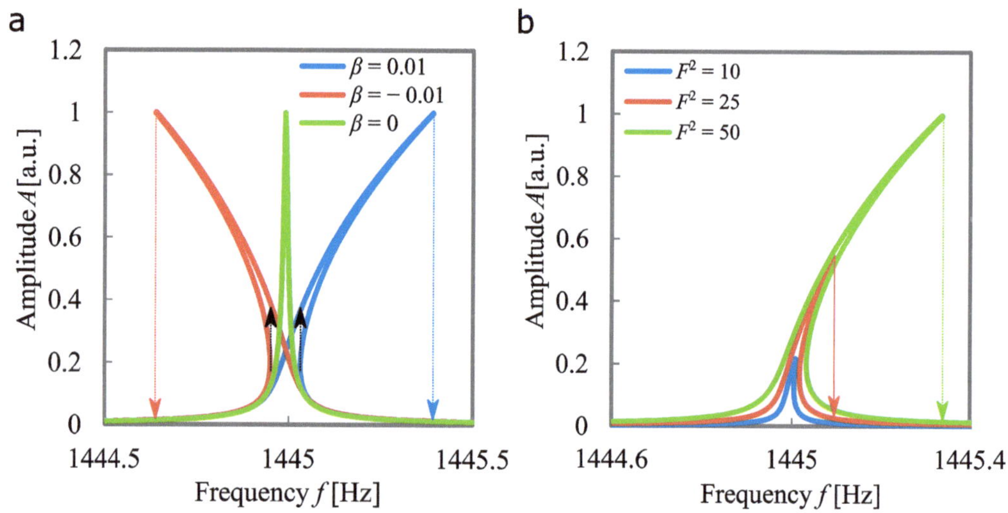

Figure 2. Numerically simulated frequency response of a Duffing oscillator for varying stiffness coefficients (**a**) and for varying actuation forces (**b**).

Contributions to the stiffness parameter for piezoelectric oscillators are discussed in the literature and consist of three parts [18–20]: geometric, inertial, and piezoelectric effects. Here, the geometric part β_{geo} has a hardening effect, whereas the inertial component β_{inertial} as well as the piezoelectric influence β_{piezo} have a softening effect:

$$\beta = \beta_{\text{geo}} - \beta_{\text{inertial}} - \beta_{\text{piezo}} \tag{4}$$

The geometrical hardening effect is caused by the rigidity of the solid-state material at large oscillation amplitudes, which leads to a stiffening of the spring. The inertial effects appear predominant at high kinetic energies of the oscillator. The inertial effects are defined by the velocity and the acceleration on the oscillator. These effects are dominating at higher modes, due to the increase of the resonance frequency as well as the decrease of the oscillation amplitude. The piezoelectric effects originate from the nonlinear nature of the deflection of piezoelectric materials when applying high electric fields.

The solution of the Duffing equation for a given frequency is not a single-valued function. When the non-physical negative values for the amplitude are excluded, the equation leads to one (stable region of the resonance curve), two (at the jumping point) or three solutions (unstable region of the resonance curve) for the amplitude response. Within the unstable region, the amplitude exhibits abrupt jumping phenomena, depending on the sweep direction. In Figure 2a, the different sweep directions are indicated by dashed arrows. This leads to a hysteresis behaviour of the amplitude. The strength of the nonlinear effect and therefore the size of the hysteresis loop is strongly dependent on the amplitude of the driving force F_{piezo}, which in turn depends on the actuation voltage. As can be seen in Figure 2b, the hysteresis gap (frequency difference between the two sweep directions) increases with increasing actuation force.

In the case of piezoelectrically driven oscillators, the actuation force F_{piezo} is given by the electrically induced stress tensor σ of the aluminium nitride (AlN) thin film on the cantilever. Thus, the mechanical stress can be written in the full Voigt matrix and vector notation as:

$$\sigma = \boldsymbol{d}^{\mathrm{T}}\boldsymbol{E}, \tag{5}$$

with the piezoelectric coefficient tensor $\boldsymbol{d}^{\mathrm{T}}$ [21], the electric field $\boldsymbol{E}$, which is given by the actuation voltage U, and the thickness of the AlN film t ($E = U/t$). In our configuration, the actuation force by the piezoelectric AlN thin film with the actuation area S can be simplified as follows:

$$F_{\text{piezo}} = d_{31}E_3\,S = d_{31}\frac{U}{t}S. \tag{6}$$

2.2. Ayela's Model

Due to the fact that the amplitude frequency response of Duffing oscillators is given as an implicit function (see Equation (3)), a higher numerical effort is necessary to extract the information from the resonance curve. Therefore, several approximation models have been established [13,22], to directly evaluate the key parameters out of the resonant behaviour. Important indicators include the shift of the peak frequencies, Δf_{up}, and Δf_{down}, as well as the peak amplitudes of the upward sweep direction A_{up}, and downward sweep direction A_{down} (see Figure 3). Ayela et al. [13] derived an approximation for the Duffing behaviour of electrostatically excited MEMS oscillators, extracting key parameters such as the linear attenuation and the nonlinear stiffness parameter of the oscillator. In the following, the major points of the model are presented and summarised in Table 1.

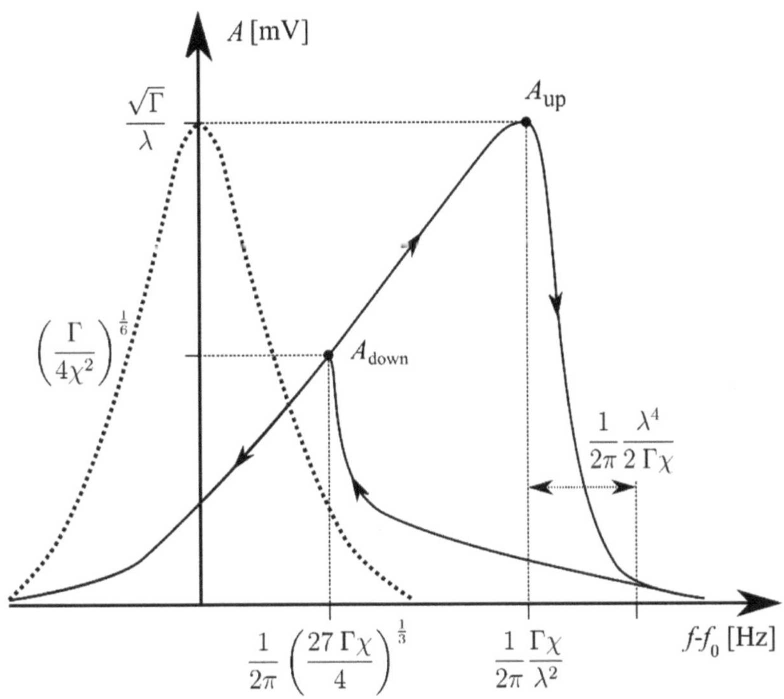

Figure 3. Amplitude versus the frequency shift for a damped harmonic oscillator (dotted line) and Duffing oscillator (solid line). The key parameters according to Ayela's model are included [9].

Table 1. Listing of important parameters for characterising the behaviour of Duffing oscillators.

Ayela's Model [9]	Amplitude [mV]	Frequency [Hz]
Upward Sweep	$A_{\text{up}} = \frac{\sqrt{\Gamma}}{\lambda}$	$\Delta f_{\text{up}} = \frac{1}{2\pi}\frac{\Gamma\chi}{\lambda^2}$
Downward Sweep	$A_{\text{down}} = \left(\frac{\Gamma}{4\chi^2}\right)^{\frac{1}{6}}$	$\Delta f_{\text{down}} = \frac{1}{2\pi}\left(\frac{27\Gamma\chi}{4}\right)^{\frac{1}{3}}$

The frequency shift $\Delta\omega$ caused by the Duffing effect is given by:

$$\omega_{\text{Duff}} = \omega_0 + \Delta\omega = \omega_0 + \frac{3}{8}\frac{\beta}{\omega_0}A^2 = \omega_0 + \chi A^2 \left[\text{s}^{-1}\right]. \tag{7}$$

The gap between A_{up} and $A \approx 0$ is given by:

$$\varepsilon = \frac{1}{2\pi}\frac{\lambda^4}{2\Gamma\chi}, \tag{8}$$

with Γ being the strength of the actuation, which can be calculated via the upward Amplitude A_{up}:

$$\Gamma = (\lambda A_{\text{up}})^2 \left[\text{m}^2\,\text{s}^{-2}\right] = \left(\frac{F_{Piezo}}{2m\omega_0}\right)^2. \tag{9}$$

Combining the equations of the frequency shift in upward and downward directions leads to an expression for the linear damping coefficient λ:

$$\lambda = \sqrt{\frac{2\,(2\pi\,\Delta f_{\text{down}})^3}{27\pi\Delta f_{\text{up}}}} = 2.4184\cdot\sqrt{\frac{(\Delta f_{\text{down}})^3}{\Delta f_{\text{up}}}}\left[\text{s}^{-1}\right]. \tag{10}$$

The nonlinear damping coefficient can either be derived from the downward sweeping amplitude A_{down} or by fitting the slope of the frequency shifts as a function of the associated amplitude Δf_{up} ($A_{\text{up}}{}^2$) or Δf_{down} ($A_{\text{down}}{}^2$) (see Equation (7)). The nonlinear damping coefficient χ is given as follows:

$$\chi = \sqrt{\frac{\Gamma}{4\,A_{\text{down}}{}^6}}\left[\text{m}^{-2}\,\text{s}^{-1}\right], \tag{11}$$

This can be related to the stiffness parameter β by using Equation (7), yielding:

$$\beta = \frac{16\pi}{3}\chi f_0 \left[\text{m}^{-2}\,\text{s}^{-2}\right]. \tag{12}$$

This model is applied to the piezoelectric MEMS oscillator presented in this work to determine the strength of the nonlinearity.

3. Experimental Section

3.1. Manufacturing

The silicon-based micro-oscillators consisted of a single bending beam, also known as a cantilever structure, with a length of 3600 µm, a width of 1800 µm, and a thickness of 10 µm (see Figure 4a). The cantilevers were covered by an AlN thin film layer for actuation and by corresponding gold electrodes for contact. The percentage of coverage of the AlN film was varied from the whole cantilever for C_100, half of the cantilever for C_50 and a third of the cantilever for C_33 (see Figure 4b). The cantilevers were fabricated by standard microtechnology processes, including lithography, sputter deposition and wet/dry etching. A highly doped p-Si wafer (ρ = 0.01 Ωcm, boron) was chosen as substrate. At first, SiO_2 was grown for electric insulation with a thickness of 120 nm in a thermal oxidation process. The SiO_2 on the bottom side served as an adhesion layer for a 550 nm Si_3N_4 film, which was deposited by PECVD. The Si_3N_4 film was used as passivation layer for the KOH etching process as part of the releasing step. Both films were structured in a single lithography step using AZ 1518 resist and a subsequent etch process in 6% HF solution. The piezoelectric AlN film with a thickness of 1100 nm was deposited by reactive sputter deposition [21]. After the AlN patterns were lithographically structured, the film was etched with 85% phosphoric acid at 80 °C. The gold electrodes were deposited via DC sputtering and etched using aqua regia at 25 °C. The cantilevers were released in a two-step process. In the first step, a thin membrane was created by time-controlled KOH etching from the backside. Afterwards, the cantilevers were released from the front side using a Bosch dry etching process. In between the two steps, the KOH cavity was filled with AZ 1518 photo resist to stop the dry etching process and prevent the cantilever from suffering mechanical damage. Finally, the cantilevers were diced and cleaned with a solution of acetone, isopropanol, water and ethanol in an ultrasonic bath.

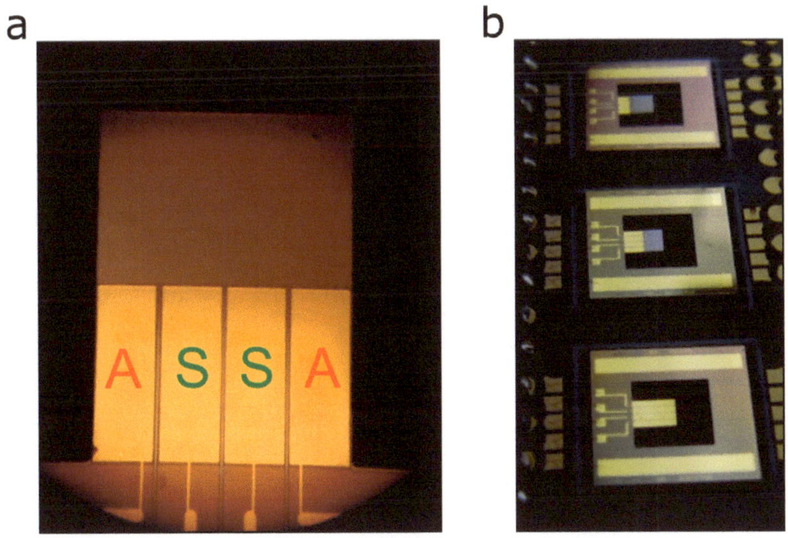

Figure 4. Top view on the cantilever structure C_50 (**a**) and measurement board with the cantilevers C_33, C_50, and C_100 (from top to the bottom) (**b**). "A" signifies the actuation electrodes, "S" signifies the sensing electrodes.

3.2. Experimental Setup

The cantilevers were glued on top of a printed circuit board (PCB) and were electrically connected by gold wire bonds. The measurement board (compare Figure 4b) was placed within a custom-built vacuum chamber providing a pressure range from high vacuum (10^{-3} mbar) up to atmospheric pressure (900 mbar). Via a feedback loop, the pressure was adjusted in a dynamic equilibrium by controlling the inlet gas flow with a mass flow controller (MFC). A pure nitrogen atmosphere was chosen to provide a clean and defined measurement environment. The pressure was measured using three pressure sensors from Pfeiffer Vacuum (CMR 261/264/362) with an accuracy of <0.2%. The cantilevers were excited by a function generator decoupled with a buffer amplifier, providing a sinusoidal signal with varying amplitude and frequency. When measuring piezoelectrically driven oscillators, a parasitic crosstalk appears which can be eliminated by a compensation circuit proposed by Qiu et al. [23]. In addition to the compensation circuit, the PCB board was equipped with a charge amplifier and with an inverting amplifier. Subsequently, the resonance curves were measured by recording the compensated and amplified frequency response of the cantilever with a lock-in amplifier (SR 5210) (see Figure 5).

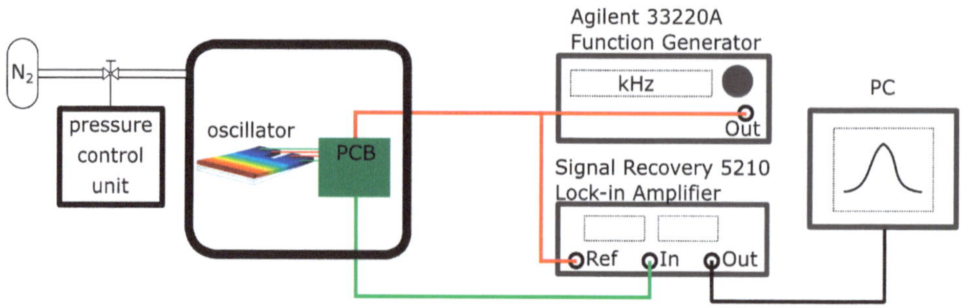

Figure 5. Schematic of the experimental setup. The vacuum chamber ranges from 10^{-3} to 10^{3} mbar.

4. Results

In order to investigate the possibility of using nonlinear Duffing MEMS oscillators for pressure sensing applications, the frequency response of the presented cantilever structures was recorded under varying ambient pressure conditions as well as excitation voltages. The hysteresis behaviour of the frequency response was recorded in upward (u) and downward (d) sweep direction around the peak frequency. For all AlN-layer coverage sizes the Duffing behaviour of the cantilevers was investigated up to a frequency of 120 kHz (limitation of the lock-in amplifier). The detected modes were associated with their shape by correlation with FEM eigenfrequency analysis and with recordings of a laser Doppler vibrometer (UHF-120 from Polytec).

The resonance curve of the micro-oscillator operating in the same higher mode can be seen in Figure 6. The results show hysteresis behaviour depending on both excitation voltage and ambient pressure. An increasing ambient pressure p leads to a decreasing hysteresis gap until it finally disappears (see Figure 6a,b). The influence of the ambient pressure on the frequency response is given by the linear damping coefficient λ, which increases with increasing pressure. In addition, the peak amplitude as well as the hysteresis loop become larger with increasing excitation voltage (see Figure 6c). These proportionalities are in accordance with the Duffing theory presented above (see Figure 2b). The dependency of the excitation voltage on the frequency response can be directly derived by taking Equation (6) into account.

The presented results show the frequency response of cantilevers with different AlN-coverage sizes (33 to 100%), operated in the same mode. A difference in the eigenfrequency appears that is mainly caused by fabrication tolerances (cantilever thickness) and, to a lesser extent, by the difference of the coverage sizes of the AlN/Au stack. The presented eigenmode is the so-called second roof-tile-shaped mode, which has been found to be advantageous for operating in heavily damped environments [24] and therefore possesses relatively large oscillation amplitudes, leading to strong Duffing effects.

Interestingly, the results in Figure 6a,b indicate a spring softening effect for 100% AlN coverage, whereas a spring hardening effect occurs for a cantilever with 33% AlN coverage, but otherwise identical geometry and mode (compare Figure 6c). According to Equation (4), the nonlinearity can be divided into a geometric part, an inertial part and a piezoelectric part [18]. The geometric nonlinearity has a spring hardening effect, whereas both inertial and piezoelectric nonlinearities have a spring softening effect. This could explain why a larger coverage area of the AlN film (C_100, Figure 6a,b) can overcome the originally dominant geometric spring hardening nonlinearity (C_33, Figure 6c) and changes it into a spring softening nonlinearity. The fundamentals behind this influence of the piezoelectric coverage area on the softening/hardening behaviour require more detailed investigation, and will be analysed in a future work. Moreover, a mode-matching optimisation of the electrode shape as well as a phase-correct actuation scheme seems promising for enhancing both sensing and actuation [25].

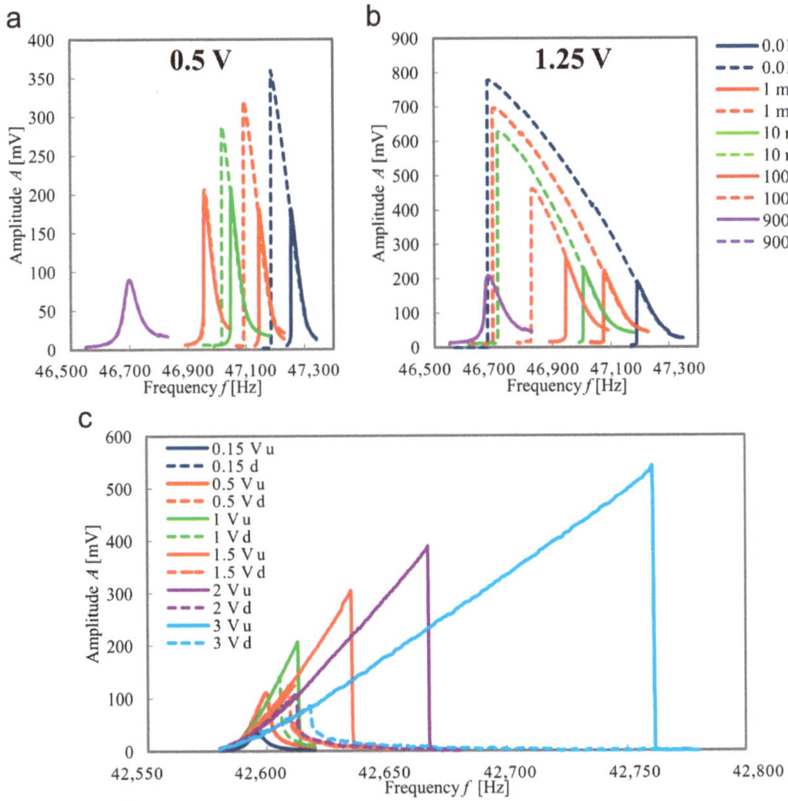

Figure 6. Amplitude frequency response in upward (u, solid) and downward (d, dotted) sweep of the 2nd roof-tile-shaped mode of sensor C_100 for different ambient pressures and for an excitation voltage of 0.5 V (**a**) and 1.25 V (**b**), respectively. Amplitude frequency response in upward (u, solid) and downward (d, dotted) sweep of the 2nd roof-tile-shaped mode of sensor C_33 for different excitation voltages at a fixed ambient pressure of 1 mbar (**c**).

To investigate the strength of the nonlinear Duffing effect, the frequency response was measured for the first four bending modes, and for the 2nd roof-tile-shaped mode. This was done at the best available vacuum level (10^{-3} mbar) to suppress the damping from the surrounding fluid. The stiffness parameter β, and the frequency shifting parameter χ (see Equations (11) and (12)) were analysed to understand the dependency of these parameters with respect to the coverage size and the mode shape. In Figure 7 the frequency shift of the peak amplitude in dependency of the squared measured amplitude is shown for both sweep directions (compare Equation (7)). The first bending mode exhibits the smallest frequency shift and biggest amplitude signal, therefore the frequency shifting parameter χ resulting from the linear fit will be the smallest. For all three coverage sizes, the frequency shifting parameter is on the same order of magnitude. The higher modes show a significantly higher frequency shift at smaller amplitudes and therefore possess a much higher value for the frequency shifting parameter. In particular, the 2nd roof-tile-shaped mode exhibits very high values, indicating a strong nonlinear Duffing effect. All results are summarised in Table 2.

The 2nd roof-tile-shaped mode (see Figure 7d,f) shows the highest value for χ within these experiments, and therefore has been chosen for further investigations under varying ambient pressure. Additionally, the spring softening behaviour of sensor C_100 operating in this mode is an interesting aspect.

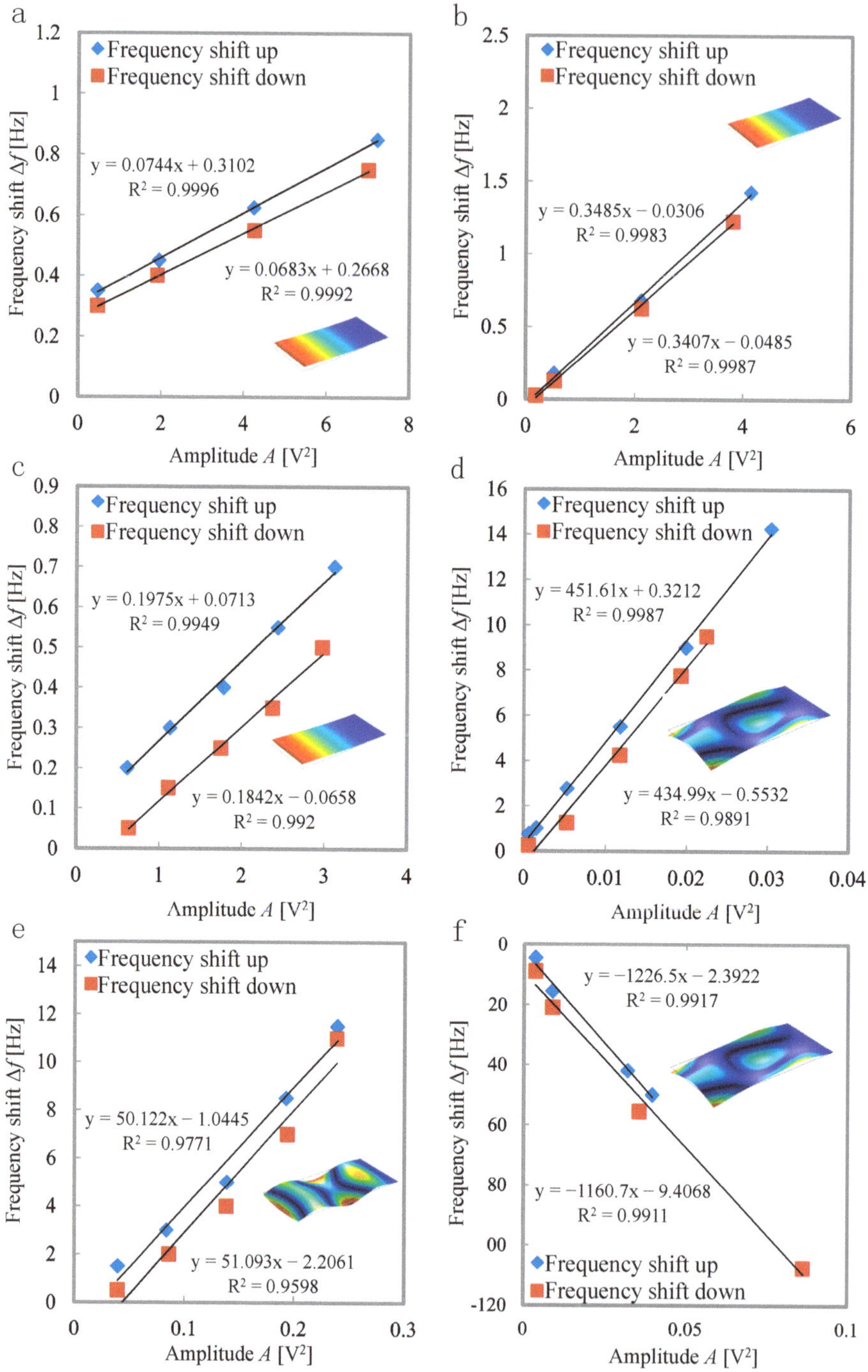

Figure 7. Frequency shift vs. squared measured amplitude for C_33 (**a**), C_50 (**b**), and C_100 (**c**) in the first bending mode, C_33 in the second roof-tile-shaped mode (**d**), C_100 in the fourth bending mode (**e**) and in the second roof-tile-shaped mode (**f**) at 0.001 mbar. The shape of the respective eigenmodes is shown in the inserts.

Table 2. Listing of determined key parameters for the presented sensors and mode shapes of Figure 7.

Sensor	Mode	Resonance Frequency f_r [Hz]	Sweep Direction	χ [m^{-2}s^{-1}]	β [m^{-2}s^{-2}]
C33	1st bending	1396.6	up	0.0744	1741
			down	0.0683	1598
C50	1st bending	1448.3	up	0.3485	8457
			down	0.3307	8025
C100	1st bending	967.9	up	0.1975	3203
			down	0.1842	2987
C100	4th bending	35543	up	50.122	29.8×10^6
			down	51.093	30.4×10^6
C33	2nd roof-tile-shape	42704.5	up	435.61	31.2×10^7
			down	434.99	31.1×10^7
C100	2nd roof-tile-shape	47355.5	up	−1226.5	-97.3×10^7
			down	−1160.7	-92.1×10^7

The possibility of cantilever structures exploiting Duffing effects for pressure sensing purposes was investigated by recording the frequency response under a defined pressure (see Figure 6). The amplitude was measured in both sweep directions for different excitation voltages. In the following, the resonance curves are analysed regarding their peak amplitudes and peak frequencies in upward and downward sweep directions. It turned out that the best correlation between measurement and ambient pressure could be achieved by the shift of the peak frequencies in upward direction in the case of hardening effects, and shift of the peak frequencies in downward direction in the case of softening effects, respectively (see Figure 8a,c). Thereby, the shift is related to the resonance frequency of the linear oscillator, which can be measured by applying a lower excitation voltage where no nonlinear effects occur.

The best results in the sense of maximum sensitivity were obtained by cantilever C_100, with an excitation voltage of 1.5 V_{pp}. A linear frequency shift of 648 Hz was measured over three pressure decades, resulting in a sensitivity of 216 Hz/decade. Normalised to the resonant frequency of 47.4 kHz, this means a frequency shift of 0.46%/decade in linear approximation. By decreasing the excitation voltage, the saturation effect can be delayed. Thus, the measurement range increases at the expense of sensitivity. Depending on the application, a reasonable compromise between sensitivity and measurement range can be adjusted by choosing an appropriate excitation voltage. Furthermore, the size of the hysteresis loop, given by the difference of the peak frequencies in both directions, is strongly affected by the ambient pressure (see Figure 8b,d). Qualitatively, this measurement shows the same behaviour as the peak frequency. The absolute values are smaller because the shift in the other sweep direction is subtracted, and therefore the sensitivity (change of measurement value with pressure) is slightly smaller. An advantage of this measurement principle could be that environmental factors such as temperature effects or long-term drift effects influencing the resonance frequency can be cancelled out more easily [26–29]. Other parameters, such as, e.g., peak amplitudes, are also affected by the ambient pressure [8], but the absolute change is relatively small and more difficult to detect. For this reason, the frequency measurement is preferable. A measurement series investigating the stability of the frequencies results in a standard deviation of 2.6 Hz (corresponding to 56 ppm), and therefore provides a reliable measure for detecting the ambient pressure.

A comparison with results obtained in previous works is shown in Table 3, ranking the performance of the presented AlN nonlinear micro-oscillator. Since the plot in Figure 8 is on a logarithmic scale, a nonlinear behaviour to the measured variable, the ambient pressure, can clearly be seen. Thus, it can be deduced that the sensitivity decreases with increasing pressure, when linearising this value for the individual pressure decades. The results indicate that the sensitivity related to the resonance frequency is exceeding the values obtained in the previous works, for nearly all pressure ranges, or else it is at least comparable. A special feature is the wide measurement range down to the high vacuum regime where most mechanical sensor principles have their limitations. In this low-pressure area, our sensor exhibits its highest sensitivity.

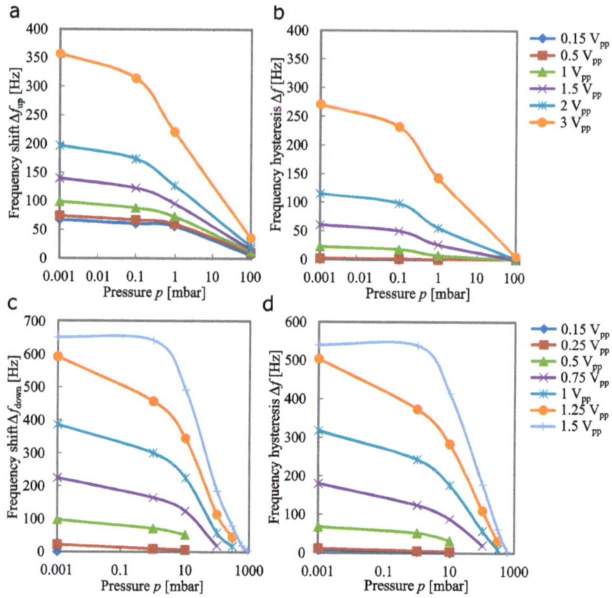

Figure 8. Frequency shift caused by the Duffing effect in dependency of the ambient pressure for the 2nd roof-tile-shaped mode of C_33 (**a**) and C_100 (**c**) for varying excitation voltages. Frequency hysteresis between the sweep directions in dependency of the ambient pressure for the 2nd roof-tile-shaped mode of C_33 (**b**), and C_100 (**d**) for varying excitation voltages.

Table 3. Comparison of the key parameters for different pressure sensing systems.

Sensor	Measurement Range [mbar]	Sensitivity [Hz/mbar]	Sensitivity [ppm/mbar]	Measurement Principle	Resonance Frequency
Zuo [30]	50–2000	16.5	0.1	AlN Contour mode resonator	140 MHz
Wang [31]	100–4000	221	0.27	AlN Contour mode resonator	820 MHz
Anderas [32]	0.1–500	360	0.4	AlN Contour mode resonator	900 MHz
Rodriguez-Madrid [33]	1000–4000	330	0.03	AlN Surface acoustic wave resonator	10.8 GHz
Han [34]	0.01–2000	1.9	35	Capacitive oscillator with piezoelectric read out	53 kHz
Shi [35]	100–1500	8	94	Capacitive oscillator with piezoelectric read out	85 kHz
This work C_100 U = 1.5 V_{pp}	1–10 10–100 100–900	17 3.4 0.23	362 72 4.9	Duffing nonlinearity	47 kHz
This work C_100 U = 1.25 V_{pp}	0.001–1 1–10 10–100 100–300	136 13 2.6 0.35	2894 277 55 7.4	Duffing nonlinearity	47 kHz

In the next steps, the gas type dependence will be investigated by characterising the sensor under different gas atmospheres as well as the temperature influence of the measurement principle. A further work which investigates the nonlinear effect of different geometries and modes to optimise the sensor response can already be found in [36].

5. Conclusions

In this study, we analysed the frequency response of a cantilever structure with different piezoelectric AlN coverage sizes. The resonance behaviour was investigated with a focus on the nonlinear Duffing effect. The strength of the nonlinearity was experimentally identified, and the Duffing parameters were determined for each AlN coverage size and for each eigenmode. Based on these results, the 2nd roof-tile-shaped mode, which was found to be the best mode according to the stiffness parameters, was selected for pressure measurements over a wide range of six decades. The quantity with the strongest correlation to the ambient pressure was found to be the frequency shift of the peak amplitude. For the 2nd roof-tile-shaped mode, a maximum sensitivity of 0.29%/mbar for the high-vacuum regime was experimentally found. In addition, the influence of the excitation voltage was shown, revealing the possibility of adjusting the sensitivity and measurement range depending on the specific application. Compared to the other measurement principles presented in the paper, the frequency shift induced by the Duffing effect of piezoelectrically actuated MEMS oscillators offers a significantly higher sensitivity, exceeding existing work, especially in the high-vacuum range.

The current state of the art for high-dynamic-range pressure sensing in vacuum chamber applications usually requires a combination of several high cost sensors (e.g., Pirani and membrane). The new principle introduced in this work, using the Duffing-effect-induced frequency shift of a micro-oscillator to obtain high-quality pressure measurements, opens up an opportunity for a new class of wide range vacuum sensors, which can be fabricated in a much more cost-effective way. Alternatively, the size of the hysteresis loop can also be exploited with the same type of sensor, giving the advantage of reducing environmental factors such as temperature effects or long-term drift. In future work, the fundamentals behind the strong Duffing effect of the selected mode need to be clarified more precisely, as well as the influence of the actuator coverage size on the spring softening or spring hardening behaviour.

Author Contributions: Investigation, T.Z. and M.S.; Data curation, T.Z.; Project administration A.A. and H.S.; Funding acquisition A.A. and H.S., Writing- original draft T.Z. and H.S.; Writing review & editing, T.Z. and H.S. All authors have read and agreed to the published version of the manuscript.

Funding: This research was funded by the German Science Foundation (DFG) within the grant SE 1425/14-1.

Institutional Review Board Statement: Not applicable.

Informed Consent Statement: Not applicable.

Data Availability Statement: Data available on request.

Acknowledgments: We thank Frank Tiefensee and Jorge Oevermann from the Fraunhofer Society for providing access to the laser vibrometer.

Conflicts of Interest: The authors declare that there is no conflict of interest.

References

1. Aikele, M.; Bauer, K.; Ficker, W.; Neubauer, F.; Prechtel, U.; Schalk, J.; Seidel, H. Resonant accelerometer with self-test. *Sens. Actuator A Phys.* **2001**, *92*, 161–167. [CrossRef]
2. Binnig, G.; Quate, C.F.; Gerber, C. Atomic force microscope. *Phys. Rev. Lett.* **1986**, *56*, 930–934. [CrossRef]
3. Chen, S.T.; Du, S.; Arroyo, E.; Jia, Y.; Seshia, A. Utilising Nonlinear air damping as a soft mechanical stopper for MEMS vibration energy harvesting. *J. Phys. Conf. Ser.* **2016**, *773*, 012098. [CrossRef]
4. Almog, R.; Zaitsev, S.; Shtempluck, O.; Buks, E. Signal amplification in a nanomechanical Duffing resonator via stochastic resonance. *Appl. Phys. Lett.* **2007**, *90*, 013508. [CrossRef]

5. Sharma, M.; Sarraf, E.H.; Baskaran, R.; Cretu, E. Parametric resonance: Amplification and damping in MEMS gyroscopes. *Sens. Actuator A Phys.* **2012**, *177*, 79–86. [CrossRef]
6. Li, Y.; Li, H.; Xiao, Y.; Cao, L.; Guo, Z. A compensation method for nonlinear vibration of silicon-micro resonant sensor. *Sensors* **2021**, *21*, 2545. [CrossRef]
7. Hasan, M.; Alsaleem, F.; Oukad, H. Novel threshold pressure sensors based on nonlinear dynamics of MEMS resonators. *J. Micromech. Microeng.* **2018**, *28*, 065007. [CrossRef]
8. Keskar, G.; Elliott, B.; Gaillard, J.; Skove, M.J.; Rao, A.M. Using electric actuation and detection of oscillations in microcantilevers for pressure measurements. *Sens. Actuator A Phys.* **2008**, *147*, 203–209. [CrossRef]
9. Kacem, N.; Arcamone, J.; Perez-Murano, F.; Hentz, S. Dynamic range enhancement of nonlinear nanomechanical resonant cantilevers for highly sensitive NEMS gas/mass sensor applications. *J. Micromech. Microeng.* **2010**, *20*, 045023. [CrossRef]
10. Green, P.L.; Worden, K.; Atallah, K.; Sims, N.D. The effect of Duffing-type non-linearities and Coulomb damping on the response of an energy harvester to random excitations. *J. Intell. Mater. Syst. Struct.* **2012**, *23*, 2039–2054. [CrossRef]
11. Millar, S.; Desmulliez, M. MEMS ultra low leak detection methods: A review. *Sens. Rev.* **2009**, *29*, 339–344. [CrossRef]
12. Martínez Rojas, J.A.; Fernández, J.L.; Sánchez Montero, R.; López Espí, P.L.; Diez-Jimenez, E. Model-based systems engineering applied to trade-off analysis of wireless power transfer technologies for implanted biomedical microdevices. *Sensors* **2021**, *21*, 3201. [CrossRef]
13. Ayela, F.; Fournier, T. An experimental study of anharmonic micromachined silicon resonators. *Meas. Sci. Technol.* **1998**, *9*, 1821–1830. [CrossRef]
14. Nabholz, U.; Heinzelmann, W.; Mehner, J.E.; Degenfeld-Schonburg, P. Amplitude- and gas pressure-dependent nonlinear damping of high-Q oscillatory MEMS micro mirrors. *J. Microelectromech. Syst.* **2018**, *27*, 383–391. [CrossRef]
15. Lifshitz, R.; Cross, M.C. Nonlinear Dynamics of Nanomechanical and Micromechanical Resonators. In *Review of Nonlinear Dynamics and Complexity*; John Wiley & Sons: Hoboken, NJ, USA, 2008.
16. Duffing, G. *Erzwungene Schwingungen bei Veränderlicher Eigenfrequenz und ihre Technische Bedeutung*; Vieweg & Sohn: Braunschweig, Germany, 1918.
17. Tajaddodianfar, F.; Yazdi, M.R.H.; Pishkenari, H.N. Nonlinear dynamics of MEMS/NEMS resonators: Analytical solution by the homotopy analysis method. *Microsyst. Technol.* **2017**, *23*, 1913–1926. [CrossRef]
18. Mahmoodi, S.N.; Jalili, N.; Daqaq, M.F. Modeling, nonlinear dynamics, and identification of a piezoelectrically actuated microcantilever sensor. *IEEE ASME Trans. Mechatron.* **2008**, *13*, 58–65. [CrossRef]
19. McHugh, K.; Dowell, E. Nonlinear responses of inextensible cantilever and free–free beams undergoing large deflections. *J. Appl. Mech.* **2018**, *85*, 051008. [CrossRef]
20. Villanueva, L.G.; Karabalin, R.B.; Matheny, M.H.; Chi, D.; Sader, J.E.; Roukes, M.L. Nonlinearity in nanomechanical cantilevers. *Phys. Rev. B* **2013**, *87*, 024304. [CrossRef]
21. Ababneh, A.; Schmid, U.; Hernando, J.; Sánchez-Rojas, J.L.; Seidel, H. The influence of sputter deposition parameters on piezoelectric and mechanical properties of AlN thin films. *Mater. Sci. Eng. B Solid State Mater Adv. Technol.* **2010**, *172*, 253–258. [CrossRef]
22. Davis, W.O. Measuring quality factor from a nonlinear frequency response with jump discontinuities. *J. Microelectromech. Syst.* **2011**, *20*, 968–975. [CrossRef]
23. Qiu, H.; Schwarz, P.; Völlm, H.; Feili, D.; Wu, X.; Seidel, H. Electrical crosstalk in two-port piezoelectric resonators and compensation solutions. *J. Micromech. Microeng.* **2013**, *23*, 045007. [CrossRef]
24. Kucera, M.; Wistrela, E.; Pfusterschmied, G.; Ruiz-Díez, V.; Manzaneque, T.; Sánchez-Rojas, J.L.; Schalk, J.; Bittner, A.; Schmid, U. Characterization of a roof tile-shaped out-of-plane vibrational mode in aluminum-nitride-actuated self-sensing micro-resonators for liquid monitoring purposes. *Appl. Phys. Lett.* **2014**, *104*, 233501. [CrossRef]
25. Sanchez-Rojas, J.L.; Hernando, J.; Donoso, A.; Bellido, J.; Manzaneque, T.; Ababneh, A.; Seidel, H.; Schmid, U. Modal optimization and filtering in piezoelectric microplate resonators. *J. Micromech. Microeng.* **2010**, *20*, 055027. [CrossRef]
26. Kim, B.; Candler, R.N.; Hopcroft, M.A.; Agarwal, M.; Park, W.T.; Kenny, T.W. Frequency stability of wafer-scale encapsulated MEMS resonators. *Sens. Actuator A Phys.* **2007**, *136*, 125–131. [CrossRef]
27. Prikhodko, I.; Trusov, A.; Shkel, A. Compensation of drifts in high-Q MEMS gyroscopes using temperature self-sensing. *Sens. Actuator A Phys.* **2013**, *201*, 517–524. [CrossRef]
28. Tanner, D.; Olsson, R., III; Parson, T.; Crouch, S.; Walraven, J.; Ohlhausen, J. Stability experiments on MEMS aluminum nitride RF resonators. *Proc. SPIE* **2010**, *7592*, 759209.
29. Defoort, M.; Taheri-Tehrani, P.; Horsley, D.A. Exploiting nonlinear amplitude-frequency dependence for temperature compensation in silicon micromechanical resonators. *Appl. Phys. Lett.* **2016**, *109*, 153502. [CrossRef]
30. Zuo, J.; Zhang, H.; Chang, Y.; Liang, J.; Pang, W.; Duan, X. Highly sensitive AlN contour-mode resonator-based pressure sensor for in-line monitoring of chemical reactions. In Proceedings of the 20th International Conference on Solid-State Sensors, Actuators and Microsystems & Eurosensors XXXIII (TRANSDUCERS & EUROSENSORS XXXIII), Berlin, Germany, 23–27 June 2019; pp. 1373–1376.
31. Wang, T.; Tang, Z.; Lin, H.; Zhan, K.; Wan, J.; Wu, S.; Gu, Y.; Luo, W.; Zhang, W. A low temperature drifting acoustic wave pressure sensor with an integrated vacuum cavity for absolute pressure sensing. *Sensors* **2020**, *20*, 1788. [CrossRef]

32. Anderås, E.; Katardjiev, I.; Yantchev, V. Lamb wave resonant pressure micro-sensor utilizing a thin-film aluminium nitride membrane. *J. Micromech. Microeng.* **2011**, *21*, 085010. [CrossRef]
33. Rodríguez-Madrid, J.G.; Iriarte, G.F.; Williams, O.A.; Calle, F. High precision pressure sensors based on SAW devices in the GHz range. *Sens. Actuator A Phys.* **2013**, *189*, 364–369. [CrossRef]
34. Han, X.; Mao, Q.; Zhao, L.; Li, X.; Li, W.; Yang, P.; Wang, Y.; Yan, X.; Wang, S.; Zhu, N.; et al. Novel resonant pressure sensor based on piezoresistive detection and symmetrical in-plane mode vibration. *Microsyst. Nanoeng.* **2020**, *6*, 95. [CrossRef]
35. Shi, X.; Lu, Y.; Xie, B.; Li, Y.; Wang, J.; Chen, D.; Chen, J. A resonant pressure microsensor based on double-ended tuning fork and electrostatic excitation/piezoresistive detection. *Sensors* **2018**, *18*, 2494. [CrossRef] [PubMed]
36. Zengerle, T.; Stopp, M.; Ababneh, A.; Seidel, H. Investigations on nonlinearities of roof-tile shape modes for pressure measurement applications. In Proceedings of the 21th International Conference on Solid-State Sensors, Actuators and Microsystems, TRANSDUCERS 21, Online, 20–25 June 2021.

actuators

MDPI

Article

Efficiency Model for Traveling Wave-Type Ultrasonic Motors Based on Contact Variables and Preload

Haoyu Sun, Hao Yin *, Jiang Liu and Xilong Zhang

School of Mechanical and Automotive Engineering, Qingdao University of Technology, Qingdao 266520, China; sunhaoyu@qut.edu.cn (H.S.); liujiang@qut.edu.cn (J.L.); zhangxilong@qut.edu.cn (X.Z.)
* Correspondence: yinhao_vehicle@qut.edu.cn; Tel.: +86-138-6485-6977

Abstract: The contact interface variables are difficult to measure for an ultrasonic motor. When the ultrasonic motor works under different preloads, the error between the traditional efficiency model and the real output is quite large. In order to solve these two problems, we propose a novel efficiency model. It takes measured preload and the feedback voltage data as the input, which may offer better accuracy and on-line ability. Firstly, the effect of the preload on the drive characteristics is investigated, and the relationship between preload and the change in motor energy input is analyzed. Secondly, a contact model based on measured preload and feedback voltage is built, providing a more accurate description of the contact variables. Finally, an efficiency model was developed with a new composite stator structure. A preload test rig for a 60 mm ultrasonic motor is built and real operating conditions are measured. The results show that the correlation coefficient of the present model is 0.991, larger than 0.925 of the conventional model. The proposed model is more consistent with the real working conditions for the motor.

Keywords: ultrasonic motors; preload; feedback voltage; contact interface; efficiency model

Citation: Sun, H.; Yin, H.; Liu, J.; Zhang, X. Efficiency Model for Traveling Wave-Type Ultrasonic Motors Based on Contact Variables and Preload. *Actuators* **2021**, *10*, 158. https://doi.org/10.3390/act10070158

Academic Editor: Jose Luis Sanchez-Rojas

Received: 3 June 2021
Accepted: 7 July 2021
Published: 12 July 2021

1. Introduction

Ultrasonic motors have the advantages of high torque density, high power density, compact structure, high positioning accuracy and fast response speed [1]. It has been broadly used in the fields of robotics [2], medical devices [3–5], microdrives [6–9] and precision drive platforms [10,11]. The output efficiency is one of the crucial performance indicators for actuators [12]. The contact interface between the stator and the rotor plays a key role in output efficiency. Besides the pressure and friction state, the motor working principle may involve a complex multi-physical field coupling process. Thus, the contact variable under different preloads has drawn many scholar's study interest.

The energy output of a motor comes from two main processes, the generation of traveling waves on the stator surface and the transfer process at the contact interface, so the prerequisite for the study of output efficiency is the analysis of the contact interface. For the motor contact interface, the contact models proposed so far have made more simplifications, many of them only consider the motion of the stator surface mass in the circumferential and axial directions of the contact interface. In order to improve the positioning accuracy elastomers of the head in the positioning control system of a hard disk drive, Shen [13] developed a three-dimensional finite element contact model to analyze the fatigue and wear processes of the friction material. Ren [14] used the finite element software ADINA to analyze the radial, circumferential and axial displacements of the contact points. Ran [15] developed a novel three-dimensional model on a cylindrical coordinate system to describe a miniature piezoelectric actuator without a tooth structure on the stator surface in order to study the drive characteristics of Micro Electro Mechanical Systems(MEMS). The finite element model is undoubtedly more accurate, but the more complex model may result in difficulties in constraint adjustment.

In order to simplify the contact model, many researchers have utilized two-dimension modes. Most of these models consist of the circumferential and axial motion of a mass point on the stator. For the transient response of the rotor of an ultrasonic motor, Tomoaki [16] developed a point contact model based on the elliptical trajectory of the stator surface mass. This model described the dynamic processes of torque and angular velocity of the motor rotor. For validation, he used a high-speed microscope to observe the elliptical trajectory and amplitude. Radi [17] derived a mathematical model to define the dynamic contact state between the stator ring and the friction body. It could be used to predict the failure source for ultrasonic motors. Renteria-Marquez [18] developed a new two-dimension contact model with the finite volume method. The calculation results and the actual data showed a good agreement. It can be seen that mathematical models also play an important role in the study of describing contact interfaces. It has been proved that these simplified models would work well in kinematic and dynamic studies for ultrasonic motors. However, for the output efficiency calculation, the relationship between the preload and the feedback voltage needs more detailed study.

A lot of work has been performed on the output efficiency. Zhang [19] changed the friction angle of the tooth structure on the stator surface, increasing the maximum efficiency of an ultrasonic motor by nearly two times. Liu [20] designed a traveling wave-type hollow ultrasonic motor to improve the output efficiency by determining the effective electromechanical coupling coefficient of the stator. The effect of the stator vibration mode, stator ring and piezoelectric sheet structure size on the effective electromechanical coupling coefficient was investigated. The experimental results of the maximum efficiency were very close to the actual data. The above studies derived mathematical models directly, and merely focus on a certain preload for a certain structure. Furthermore, the supports by the real-time measurement are not mentioned too. The literature [21] establishes a basic idea for efficiency calculations, where many of the parameters need to be measured experimentally. Contact parameters for different preload conditions are measured experimentally with greater inconvenience. The method given in this paper will improve the accuracy of the contact parameters for different preload conditions by identifying the parameters for the easily measured preload and feedback voltage, which can improve the accuracy of the output efficiency prediction of the traveling wave type ultrasonic motor.

In order to model the contact state, we identify the conversion factor by the preload and feedback voltage, rather than theoretically derivation. The load and the voltage are measured in real-time, which improves the accuracy of the contact model. This method allows us to collect the contact variables at different preloads on-line. The contact variables can be used directly in the efficiency model.

The paper is organized as follows: Section 2 describes the theoretical process of the contact and efficiency models, Section 3 shows the ultrasonic motor preload experiments, simulation results and experimental results. Section 4 gives the conclusions of the paper.

2. Motor Models

In this section, the contact and efficiency models of ultrasonic motors are investigated. For the contact model, variables, such as the amplitude and contact range of the contact interface, are relatively difficult to measure during motor operation. The traditional two or three-dimensional geometric models established in the analytical method can also cause distortions due to excessive simplification. The literature [22] gives a parameter identification method for preload and feedback voltage. It provides a simple prediction for the contact variables under different drive frequencies. Thus we expanded this identification philosophy and proposed a new approach for the preload study. The new method is based on the preload and feedback voltage data measured on-line. The conversion factor is computed by preload and feedback voltage under different preloads. The efficiency model uses the stator ring and the piezoelectric sheet as a composite structure to calculate the damping loss, the real-time contact state will be directly used as a parameter in the efficiency model, which distinguishes the new model from others.

2.1. Contact Model

The contact interface has three variables that need to be analyzed. The first is the contact length x_0, which is the circumferential distance between the two ends of the contact zone. A contact zone is a set for all the contact points in a single wave crest. This distance would change with the wave traveling so it may be written as $x_0(t)$. Similarly, we can define the second parameter. The drive length x_r is the circumferential distance between the two equivalence speed points and can be written as $x_r(t)$. The equivalence speed point is the dividing point between the drive zone and hindrance zone. On these two points, the friction surface micro mass and the overall rotor have the same circumferential speed. The last one W is the amplitude of the stator surface mass. Figure 1 shows a simplified diagram of the stator–rotor contact interface.

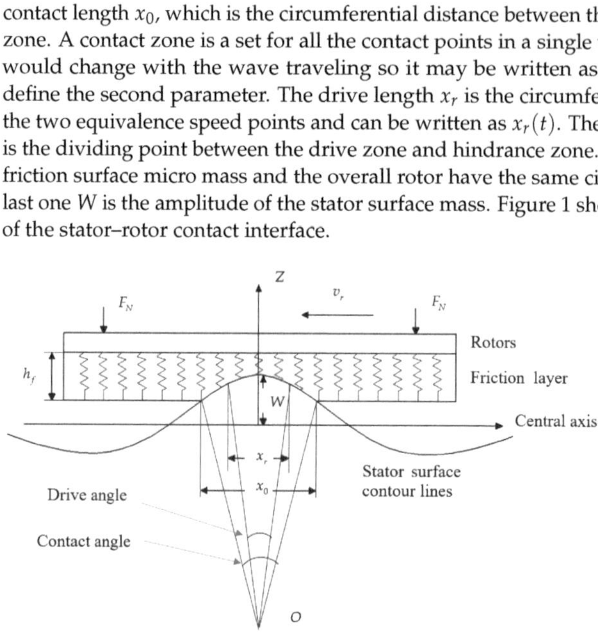

Figure 1. Model drawing of a wire spring at the stator–rotor contact interface.

The relationship between the feedback voltage and amplitude can be expressed as follows [23]:

$$V_f = K_{WV}(W\sin(\omega t) + W\cos(\omega t)) = \frac{\sqrt{2}}{2}K_{WV}W\cos\left(\omega t + \frac{\pi}{4}\right) \tag{1}$$

where V_f is the feedback voltage; K_{WV} is the conversion factor of the feedback voltage to amplitude.

The angular speed coefficient ϖ could be calculated by the modal mass m_s and the modal stiffness k_s as follows. The relative damper ratio can be written as the function of these two parameters too.

$$\varpi = \omega/\omega_n = \omega\sqrt{\frac{m_s}{k_s}} \tag{2}$$

$$\xi = c_s/\left(2\sqrt{m_s k_s}\right) \tag{3}$$

where ω_n is the angular velocity at the resonant frequency with zero preload; c_s is the modal damping of the stator.

The equivalent stiffness factor of the friction layer k_f is:

$$k_f = E_f b_f / h_f \tag{4}$$

where E_f, b_f and h_f are the modulus of elasticity, radial contact width and thickness of the friction layer.

When preload is applied, the relationship between contact distance and preload F_N can be expressed as:

$$F_N = 2n\int_0^{x_0} k_f W(\cos(kx) - \cos(kx_0))dx \tag{5}$$

where contact length is $x_0 < \lambda/4$.

The second-order dynamics model of the motor can be expressed as [24]:

$$m_s\ddot{W} + c_s\dot{W} + k_sW = N_1V_d - F_N \tag{6}$$

where V_d is the drive voltage.

Apply Equation (5) into Equation (6), which gives:

$$m_s\ddot{W} + c_s\dot{W} + k_sW = N_1V_d - N_{21}\left(2nk_fW/k\right)[\sin(kx_0) - kx_0\cos(kx_0)] - N_{22}F_N \tag{7}$$

Take $k_s{}'$ as the additional stiffness of the stator.

$$k_s{}' = N_{21}\left(2nk_f/k\right)[\sin(kx_0) - kx_0\cos(kx_0)] \tag{8}$$

The following equation is obtained:

$$F_N = \frac{(N_1V_d - N_{22}F_N)/k_s}{\sqrt{(1 + k_s'/k_s - \varpi^2)^2 + (2\xi\varpi)^2}}\frac{2nk_f}{k}[\sin(kx_0) - kx_0\cos(kx_0)] \tag{9}$$

where N_1 is the electromechanical coupling coefficient; N_{22} is the force coefficient affecting the amplitude.

When the drive voltage, drive frequency and preload are given, the contact length can be obtained by Equation (9).

The output torque can be calculated from the contact distance as follows:

$$\chi(x) = \sin(kx) - kx\cos(kx_0) \tag{10}$$

$$M_T = \frac{r\mu F_N}{\chi(x_0)}[2\chi(x_r) - \chi(x_0)] \tag{11}$$

where r is the rotor radius; μ is the coefficient of friction.

When the preload and the contact distance are certain, the output torque is zero and the drive length can be obtained [25].

The tangential velocity v_s at the stator surface can be expressed as:

$$v_s(x) = kWh_0\omega\cos(kx) \tag{12}$$

where h_0 is the distance from the stator tooth surface to the neutral plane.

$$v_r = v_s(x_r) \tag{13}$$

When a certain amount of preload is applied, the amplitude can be expressed as:

$$W = \frac{(N_1V_d - N_{22}F_N)/k_s}{\sqrt{(1 + k_s'/k_s - \varpi^2)^2 + (2\xi\varpi)^2}} \tag{14}$$

2.2. Efficiency Model

Ultrasonic motors generate various forms of energy losses during operation. The losses include the piezoelectric sheet wear, the damping loss due to stator vibration, and the friction loss at the contact interface. In the contact model, the contact angle is derived from the measured preload and feedback voltage, and the contact angle will be used directly as input parameters to the efficiency model.

The loss in piezoelectric sheets mainly includes dielectric loss, damping loss and electromechanical loss, in which the dielectric loss accounts for the largest proportion. As the piezoelectric sheet is pasted on the stator ring, the piezoelectric ceramic and stator ring can be used as a composite structure of the stator when considering the damping loss. The electromechanical loss may be neglected due to their small values.

The tangent to the dielectric loss angle δ can be expressed as the following equation:

$$\tan\delta = \frac{1}{\omega C R_n} \tag{15}$$

where C is the static capacitance and R_n is the equivalent resistance of the dielectric.

The dielectric loss over one drive voltage cycle can be expressed as:

$$W_1 = \frac{V_d^2 \sin\delta}{R_d} T \tag{16}$$

where R_d is the load-bearing equivalent resistance; T is the drive voltage period.

The piezoelectric sheet generates damping losses under high-frequency vibrations and the stator ring undergoes forced vibrations under the excitation of the piezoelectric sheet. Here the stator ring and piezoelectric sheet are used as a composite structure. The damping loss of the composite structure can be expressed as the following equation.

$$W_2 = k\pi \int_0^{2\pi r/k} \left(E_{eq} I_e \eta_e + E_p I_p \eta_p\right) \left(\frac{\partial^2 u}{\partial x^2}\right)^2 dx = 8\pi^5 W^2 k^4 \left(E_{eq} I_e \eta_e + E_p I_p \eta_p\right) / (2\pi r)^3 \tag{17}$$

where E_{eq} and E_p are the equivalent Young's modulus of the stator ring and piezoelectric sheet respectively; I_e and I_p are the cross-sectional moments of inertia of the two respectively; η_e and η_p are the damping coefficients of the two respectively. The variable μ of the second-order partial is expressed as the position on the stator structure.

The diagram below shows the schematic of piezoelectric composite structure.

In Figure 2, the width of the stator ring and stator ring is b_e. The height of the stator teeth is h_t. The height of the stator ring plus stator teeth is h_e. The stator height is h and the width of the piezoelectric ring is b_p.

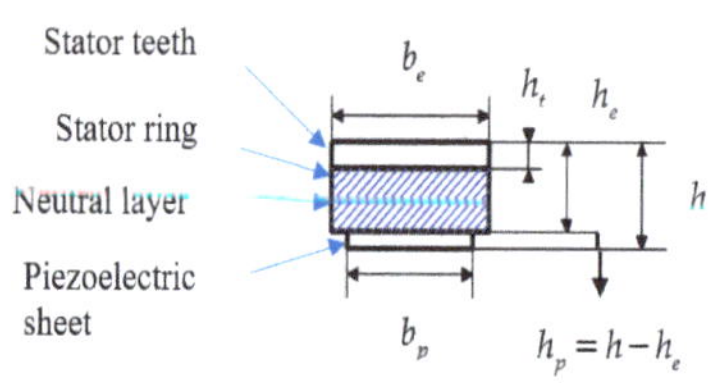

Figure 2. Schematic of piezoelectric composite structure.

The cross-sectional moment of inertia of the stator ring I_e and piezoelectric sheet I_p can be expressed as:

$$I_e = b_e \frac{D^3 + (h_e - D)^3}{3} \tag{18}$$

$$I_p = b_p \frac{(h - D)^3 + (h_e - D)^3}{12} \tag{19}$$

$$h = h_e + h_p \tag{20}$$

$$D = \frac{E_{eq} h_e^2 b_e + 2E_p h_p h_e b_p + E_p h_p^2 b_p}{2\left(E_{eq} h_e b_e + E_p h_p b_p\right)} \tag{21}$$

where D is the distance from the upper surface of the stator ring to the neutral layer. The neutral layer is illustrated in Figure 2.

Figure 3 illustrates the parameters of the contact model during a traveling wave period. The contact model gives a radian φ from the contact length x_0, which is calculated on the measured preload and feedback voltage in Section 2.1. ϕ is half of the range except for the contact zone. Then, we can get $\varphi = \pi - 2\phi$.

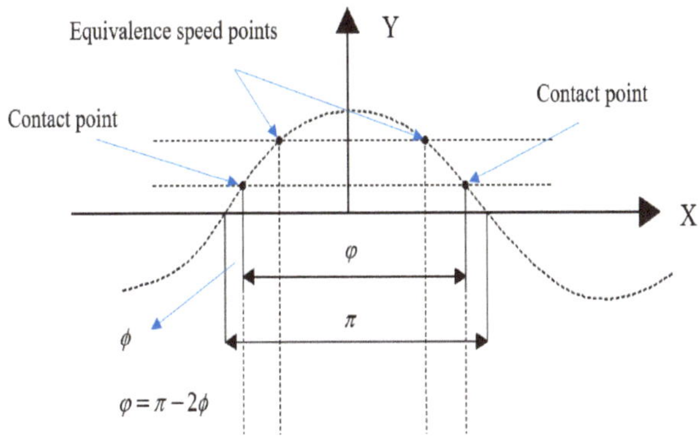

Figure 3. Schematic diagram of the contact model.

The frictional energy loss at the contact interface during a drive voltage cycle can be expressed as:

$$\begin{aligned} W_3 &= k\tfrac{\lambda}{\omega}\int_{\phi}^{\pi-\phi} \varepsilon\Delta F(v - v_m)^2 d\theta = k\tfrac{\lambda}{\omega}\int_{\phi}^{\pi-\phi} \varepsilon\Delta F(v_{s0}\sin\theta - v_m)^2 d\theta \\ &= k\varepsilon WH\tfrac{2\pi}{\omega}\left[M_1 v_m^2 - M_2 v_{s0} v_m + M_3 \tfrac{v_{s0}^2}{2}\right] \end{aligned} \tag{22}$$

where ε is the proportionality constant reflecting the magnitude of the frictional force; v_{s0} is the tangential velocity of the wave crest mass on the surface of the stator traveling wave. v_m is the rotor speed. v is the velocity of the surface mass of the traveling wave.

The four coefficients in Equation (21) can be expressed as:

$$H = k_f / k \tag{23}$$

$$M_1 = 2\sin\frac{\varphi}{2} - \varphi\cos\frac{\varphi}{2} \tag{24}$$

$$M_2 = \varphi - \sin\varphi \tag{25}$$

$$M_3 = \frac{1}{3}\sin\frac{3}{2}\varphi + 3\sin\frac{\varphi}{2} - (\varphi + \sin\varphi)\cos\frac{\varphi}{2} \tag{26}$$

The output energy of the motor during one drive voltage cycle is:

$$W_{\text{out}} = \int_t^{t+T} T_{\text{load}}\omega_m dt \tag{27}$$

where T_{load} is the load torque; ω_n is the rotor angular speed.

The output efficiency of the model was obtained as:

$$\eta_f = \frac{W_{\text{out}}}{W_{\text{out}} + W_1 + W_2 + W_3} \tag{28}$$

It is quite difficult to measure the three energy losses W_1, W_2 and W_3. Thus, we count the efficiency directly by the ratio between the input and output energies. The input energy of the motor during one drive voltage cycle is:

$$W_{in} = \int_t^{t+T} I^T \cdot V_d dt \tag{29}$$

The experimentally measured output efficiency η_m of the motor can be expressed as

$$\eta_m = W_{out}/W_{in} \tag{30}$$

3. Simulation and Experimentation

In order to verify the above model, we build a simple test rig for an ultrasonic motor. As shown in Figure 4. The TRUM-60 was driven by an ultrasonic motor driver. The preload was provided by a servo-electric cylinder (DDA-40, Dingying Intelligent Equipment Co., Ltd., Shenzhen, China). The servo-electric cylinder is driven by an actuator. A pressure transducer (TSC-1000, Toledo Corporation, Shanghai, China) sensed the preload and generated a force signal which was displayed by a digital display device. Torque sensor (CYT-303, Tianyu Hengchuang Sensor Technology Co., Ltd., Beijing, China) with data acquisition module (PXI-1031, NI, Austin, TX, USA) can record motor speed, torque and power signals. The tension controller (SC-1K, Lanling Mechanical and Electrical Technology Co., Ltd., Jiangsu, China) provides a stable current input to the magnetic powder brake (FKG-10, Lanling Mechanical and Electrical Technology Co., Ltd., Jiangsu, China). The magnetic powder brake provided a simulated load for the TRUM-60. The digital oscilloscope is used to measure the electrical signal of the ultrasonic motor.

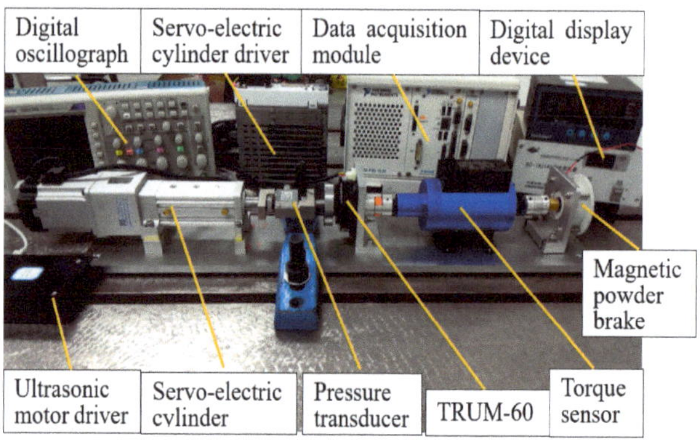

Figure 4. Diagram of ultrasonic motor experimental device.

This section analyses the drive characteristics and simulates the contact and output efficiency models, and verifies the validity of the output efficiency model through experiments.

3.1. Drive Characteristics

This section investigates the drive characteristics of the motor, using experiments to measure the effect of different preload on the two-phase drive voltage and input current. The results are shown in Figure 5.

The drive frequency is 42 kHz. As can be seen from Figure 5a, when the motor is at no load and the preload increases from 50 N to 300 N, the two-phase drive voltage amplitude gradually decreases, but the magnitude of the decrease is not obvious and is reflected in the Lissajous curve, where the enclosed zone gradually decreases and reaches a minimum at 350 N. When the preload reaches 350 N, the A-phase drive voltage amplitude decreases to 110 V and 112 V in the opposite directions. The B-phase drive voltage amplitude decreases to 101 V and 96 V in the opposite directions. The four voltage amplitudes have an average value of 106 V, which is 35 V less than the average value at 50 N. As the preload continues to increase, the two-phase voltage amplitude begins to increase again. Too much voltage difference between phase A and phase B will damage the amplitude stability of the traveling waves. If this situation happened, the stator surface mass of the elliptical motion trajectory would distort, causing unstable operations due to a poor contact state. Figure 5b shows that when the preload is 350 N and drive frequency is 42 kHz, the amplitude of the drive voltage increases as the load increases. Therefore, it can be summarized from the above two figures that, for the drive voltage without any load, an increase in preload will result first in a decrease and then an increase. When the motor has a certain preload, the drive voltage will increase with the load.

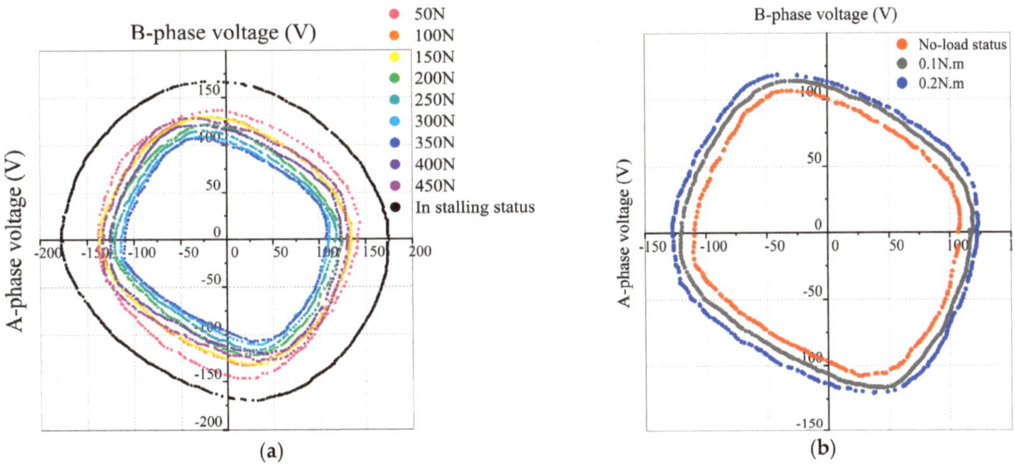

Figure 5. Variation of two-phase drive voltage with preload: (**a**) No-load status; (**b**) Different load states at a preload of 350 N.

The drive frequency is 42 kHz. The load is zero. Figure 6 shows that as the preload increases, the current increases exponentially. The minimum current is 0.22 A at 50 N preload, When the preload is 400 N the maximum current is 0.6 A.

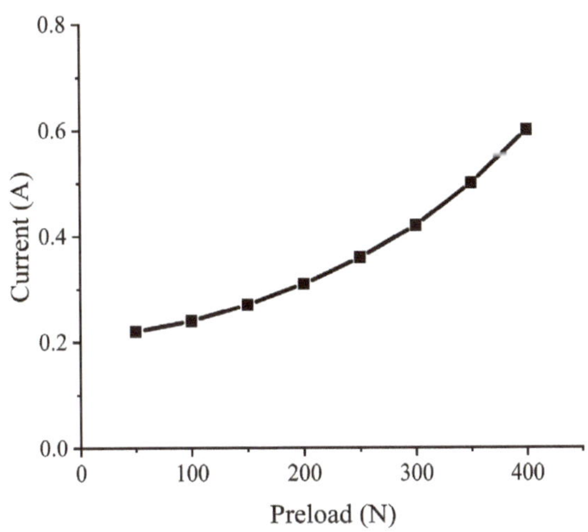

Figure 6. Current variation at different preloads.

The input power is equal to the drive voltage multiplied by the input current, both the drive voltage and the input current vary with preload. Therefore, an increase in the preload or in the load will result in an increasing the energy input to the motor.

3.2. Contact Characteristics

The no-load speed and feedback voltage were collected at different preloads. Then we fitted these data with reference to the literature [22]. The motor identification parameters are shown in Table 1. The relevant performances were measured and listed in Table 2.

Table 1. Motor identification parameters.

Motor Identification Parameters	Value
$h_0(m)$	4.0×10^{-3}
$E_f(N/m^2)$	4.3×10^{9}
b_f(m)	3.2×10^{-3}
$m_s(kg)$	0.55×10^{-3}
$c_s(Ns/m)$	8.8×10^{-2}
$k_s(N/m)$	345
μ	0.24
$r(m)$	3.0×10^{-2}
$tan\delta$	3.5×10^{-3}

Table 2. Contact model parameter.

Preload (N)	No-Load Speed (rpm)	Feedback Voltage (V)	Conversion Factor
100	106	38	3.07×10^7
200	144	46	3.24×10^7
300	138	40	3.11×10^7
400	89	34	2.86×10^7

Putting the above conversion factors into the contact model, we can obtain the variation of the contact interface within a wave lengths. Five characterristc curves, including the contact angle and length, the drive angle and the length, and the amplititude, are shown in Figure 7.

In Figure 7a, the drive frequency is 42 kHz and the drive voltage is 150 V. The contact length increases rapidly as the preload increases from zero. As the preload increases to 50 N, the growth of the contact length begins to slow down. The growth rate is approximately 1.8×10^{-3} mm/N until the contact length gradually reaches a quarter of the wavelength. The trend of the contact angle is similar to that of the contact length.

In Figure 7b, the trend of the drive angle and length are similar to that of the contact length. However, they are much less than the contact angle and length.

Figure 7c shows the relationship between amplitude and preload. The amplitude increases slightly when the preload is 0 N–100 N and decreases rapidly when it is greater than 100 N until the preload increases to 300 N. After that, the amplitude begins to decrease more slowly.

The above shows that there is a drive and hysteresis zone at the contact interface. The size of these drive two zones is non-linear, and the tooth structure on the stator will directly exacerbate this non-linear effect.

3.3. Efficiency Features

The output efficiency model was simulated to obtain three energy losses in one traveling wave cycle. The simulation curves are shown in Figure 8.

From Figure 8a, the dielectric loss is defined by the drive voltage and the electrical parameters. The drive voltage is influenced by preload. Based on the results in Section 3.1, we can find that the dielectric loss follows a similar trend to the drive voltage with preload. Dielectric loss decreases and then increases with preload. Losses are minimized when the pre-pressure reaches 350 N. Figure 8b shows that the damping loss of the composite stator follows a similar trend to the variation of amplitude with preload. It keeps increasing in the range 0–100 N. When the preload is 100 N–300 N, the damping loss decreases rapidly. After that, the decrease slows down and W_2 tends to 5×10^{-3} J. This is because the energy loss in material bending is proportional to the square of the vibration amplitude, the bending stiffness and bending angle, according to Equation (17). Figure 8c shows that the frictional losses will increase with increasing preload.

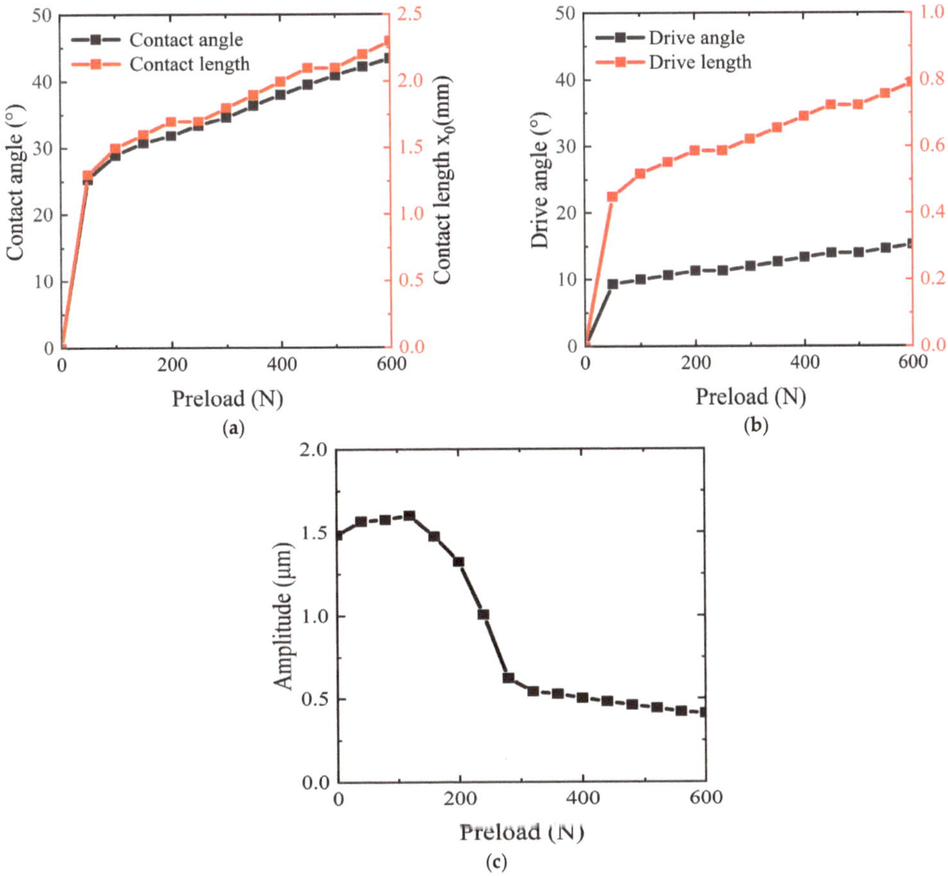

Figure 7. Contact interface parameters for no-load conditions at different preloads: (**a**) Contact length and contact angle; (**b**) Drive length and drive angle; (**c**) Amplitude.

Figure 9a shows the efficiency's overall map with the preload and output torque. The driving frequency is 42 kHz. The drive voltage is 150 V. The output efficiency's 3D surface basically maintains the trend of increasing and then decreasing. We choose a typical 0.2 N.m torque and illustrated the simulation and experimental result in Figure 9b. A simulation curve with the traditional model is added for comparison. The driving frequency is 42 kHz. The drive voltage is 150 V. The experimental curve is below the simulation curve of the new model for two main reasons: firstly, the efficiency model is built with more simplifications, simplifying the stator ring with teeth to a simply supported beam, resulting in smaller stator damping losses than the actual losses; compared to the 3D contact model, the contact interface is simplified to a 2D contact model, and no radial sliding losses are considered. Secondly, only three relatively large energy losses are considered in the efficiency model, and other loss types are not accounted into simulations.

It can be seen from Figure 9b that the efficiency model in this paper has a higher correlation than the traditional efficiency model [26]. The Pearson correlation coefficient between the traditional efficiency model and the real working condition of the motor is 0.925, while for the new model it is 0.991.

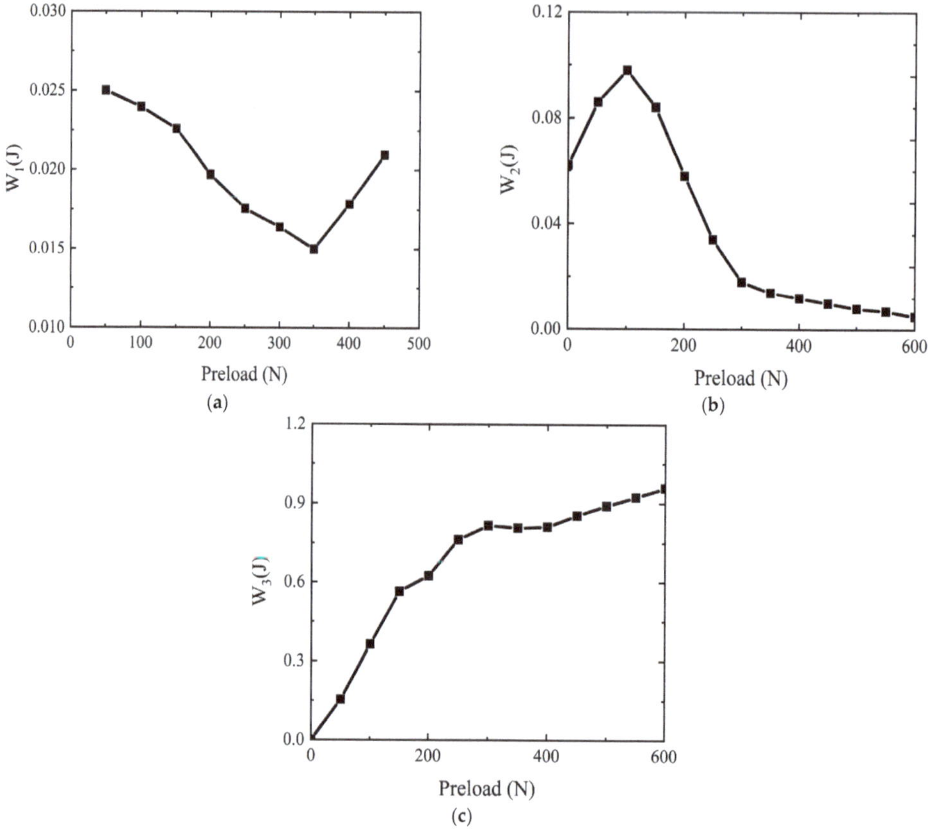

Figure 8. Three types of energy loss versus preload for one travelling wave period: (**a**) Dielectric loss W_1; (**b**) Damping loss W_2; (**c**) Frictional loss W_3.

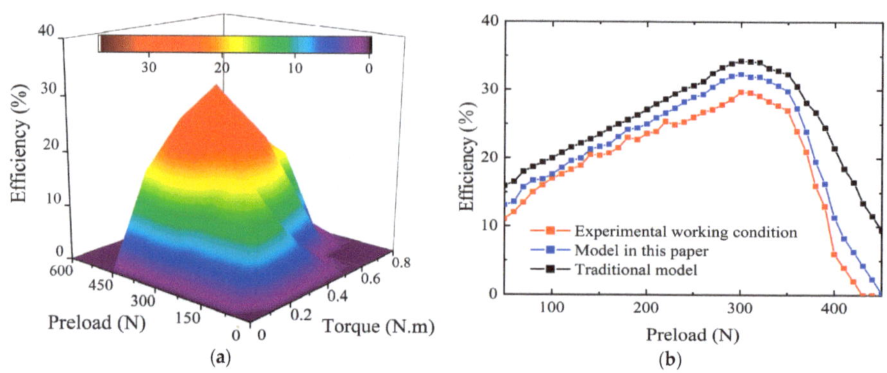

Figure 9. Motor output efficiency: (**a**) Experimental results of preload–load torque–output efficiency; (**b**) Comparison of output efficiency for a load torque of 0.2 N.m.

4. Conclusions

In this paper, a contact model and efficiency model based on measured preload and feedback voltages are presented. The contact interface parameters and output efficiency of the ultrasonic motor are analyzed by measuring the preload and feedback voltage in real-time. Through the simulations and test experiments, the following conclusions are obtained:

(1) The proposed new model could offer more accurate interface variables by real-time identifications. The composite stator structure allows us to calculate the output performance more easily. Comparing with the traditional model, the Pearson correlation coefficient with real operating conditions increases from 0.925 to 0.991. A higher Pearson coefficient may lead to a better quality for simulations.

(2) The simulation results of the contact model show that the proportion of the drive zone to the contact zone gradually decreases with the preload. The stator surface mass vertical amplitude first increases and then decreases. The amplitude keeps a larger value when the preload is within 200 N. The maximum amplitude is 1.6 microns when the preload is 120 N.

(3) By the motor drive characteristics test, we found that when the motor is in the no-load state, the drive voltage first decreases and then increases with the preload increases. When the load increases from zero to 0.2 N.m, the drive voltage gradually increases with a constant preload. For a constant load, the input current and the input energy will also increase with the preload.

(4) The friction loss, increasing with the preload, accounts for the largest proportion of the three kinds of losses. As to the stator damping loss, the amplitude of the stator surface mass plays a very important role, which reaches its peak when the preload is about 100 N.

Author Contributions: Conceptualization, H.S. and H.Y.; methodology, H.S.; software, H.S.; validation, H.S., H.Y. and J.L.; formal analysis, H.Y.; writing—original draft preparation, H.S.; writing—review and editing, H.S.; supervision, X.Z.; project administration, H.Y.; funding acquisition, J.L. All authors have read and agreed to the published version of the manuscript.

Funding: This research was funded by National Natural Science Foundation of China (51575288) and Natural Science Foundation of Shandong Province General Project (ZR2019MEE089).

Conflicts of Interest: The authors declare no conflict of interest.

References

1. Pulskamp, J.S.; Polcawich, R.G.; Rudy, R.Q.; Bedair, S.S.; Proie, R.M.; Ivanov, T.; Smith, G.L. Piezoelectric PZT MEMS technologies for small-scale robotics and RF applications. *MRS Bull.* **2012**, *37*, 1062–1070. [CrossRef]
2. Feng, Z.; Liang, W.; Ling, J.; Xiao, X.; Tan, K.K.; Lee, T.H. Integral terminal sliding-mode-based adaptive integral backstepping control for precision motion of a piezoelectric ultrasonic motor. *Mech. Syst. Signal Process.* **2020**, *144*, 106856. [CrossRef]
3. Takemura, K.; Park, S.; Maeno, T. Control of multi-dof ultrasonic actuator for dexterous surgical instrument. *J. Sound Vib.* **2008**, *311*, 652–666. [CrossRef]
4. Lau, J.Y.; Liang, W.; Liaw, H.C.; Tan, K.K. Sliding Mode Disturbance Observer-based Motion Control for a Piezoelectric Actuator-based Surgical Device. *Asian J. Control.* **2018**, *20*, 1194–1203. [CrossRef]
5. Wang, W.; Castro, L.A.; Hoyos, M.; Mallouk, T. Autonomous Motion of Metallic Microrods Propelled by Ultrasound. *ACS Nano* **2012**, *6*, 6122–6132. [CrossRef]
6. Smith, G.L.; Pulskamp, J.S.; Sanchez, L.M.; Potrepka, D.; Proie, R.M.; Ivanov, T.G.; Rudy, R.Q.; Nothwang, W.D.; Bedair, S.S.; Meyer, C.D.; et al. PZT-Based Piezoelectric MEMS Technology. *J. Am. Ceram. Soc.* **2012**, *95*, 1777–1792. [CrossRef]
7. Li, J.; Zhao, H.; Qu, H.; Cui, T.; Fu, L.; Huang, H.; Ren, L.; Fan, Z. A piezoelectric-driven rotary actuator by means of inchworm motion. *Sens. Actuators A Phys.* **2013**, *194*, 269–276. [CrossRef]
8. Mashimo, T. Micro ultrasonic motor using a one cubic millimeter stator. *Sens. Actuators A Phys.* **2014**, *213*, 102–107. [CrossRef]
9. Ho, S.-T.; Jan, S.-J. A piezoelectric motor for precision positioning applications. *Precis. Eng.* **2016**, *43*, 285–293. [CrossRef]
10. Mashimo, T.; Iduhara, S.; Arai, S.; Zhang, Z.; Oku, H. High-Speed Visual Feedback Control of Miniature Rotating Mirror System Using a Micro Ultrasonic Motor. *IEEE Access* **2019**, *8*, 38546–38553. [CrossRef]
11. Mashimo, T. Miniature preload mechanisms for a micro ultrasonic motor. *Sens. Actuators A Phys.* **2017**, *257*, 106–112. [CrossRef]

12. Storck, H.; Wallaschek, J. The effect of tangential elasticity of the contact layer between stator and rotor in travelling wave ultrasonic motors. *Int. J. Non-Linear Mech.* **2003**, *38*, 143–159. [CrossRef]
13. Shen, S.; Lee, H.P.; Lim, S.P.; Ong, C.J. Three-dimensional finite element analysis of interfacial delamination in traveling wave ultrasonic motors. *Int. J. Damage Mech.* **2014**, *23*, 964–978. [CrossRef]
14. Ren, W.; Yang, L.; Ma, C.; Li, X.; Zhang, J. Output performance simulation and contact analysis of traveling wave rotary ultrasonic motor based on ADINA. *Comput. Struct.* **2019**, *216*, 15–25. [CrossRef]
15. Ran, L.; Zhou, W.; He, J.; Zhan, L.; Chen, Q.; Yu, H.; Peng, B. A novel three-dimensional contact model of piezoelectric traveling wave ultrasonic micromotor. *Smart Mater. Struct.* **2020**, *29*, 075016. [CrossRef]
16. Mashimo, T.; Terashima, K. Dynamic analysis of an ultrasonic motor using point contact model. *Sens. Actuators A Phys.* **2015**, *233*, 15–21. [CrossRef]
17. Radi, B.; El Hami, A. The study of the dynamic contact in ultrasonic motor. *Appl. Math. Model.* **2010**, *34*, 3767–3777. [CrossRef]
18. Renteria-Marquez, I.; Tseng, B.T.L. A novel contact model of piezoelectric traveling wave rotary ultrasonic motors with the finite volume method. *Ultrasonics* **2018**, *90*, 5–17. [CrossRef]
19. Zhang, J.; Yang, L.; Ma, C.; Ren, W.; Zhao, C.; Wang, F. Improving efficiency of traveling wave rotary ultrasonic motor by optimizing stator. *Rev. Sci. Instrum.* **2019**, *90*, 056104. [CrossRef]
20. Liu, J.; Niu, R.; Zhu, H.; Zhao, C. Improving the efficiency of a hollow ultrasonic motor by optimizing the stator's effective electromechanical coupling coefficient. *Rev. Sci. Instrum.* **2020**, *91*, 016104. [CrossRef]
21. Wang, G.Q. Research on Several Key Problems of Traveling-wave Type Ultrasonic Motor. Ph.D. Thesis, Zhejiang University, Hangzhou, China, March 2006.
22. Li, S.; Li, D.; Yang, M.; Cao, W. Parameters identification and contact analysis of traveling wave ultrasonic motor based on measured force and feedback voltage. *Sens. Actuators A Phys.* **2018**, *284*, 201–208. [CrossRef]
23. Giraud, F.; Lemaire-Semail, B.; Aragones, J.; Robineau, J.; Audren, J.-T. Stability Analysis of an Ultrasonic Motor for a New Wave Amplitude Control. In Proceedings of the 2007 IEEE Industry Applications Annual Meeting, New Orleans, LA, USA, 23–27 September 2007; Volume 45, pp. 1343–1350. [CrossRef]
24. Zhu, M. Contact analysis and mathematical modeling of traveling wave ultrasonic motors. *IEEE Trans. Ultrason. Ferroelectr. Freq. Control.* **2004**, *51*, 668–679. [CrossRef] [PubMed]
25. Guo, J.; Gong, S.; Guo, H.; Liu, X.; Ji, K. Force transfer model and characteristics of hybrid transducer type ultrasonic motors. *IEEE Trans. Ultrason. Ferroelectr. Freq. Control.* **2004**, *51*, 387–395. [CrossRef] [PubMed]
26. Dong, Z.P. Traveling Ultrasonic Motors baesd on Effective Elliptic Motion and Structural Force Factor. Ph.D. Thesis, Hefei University of Technology, Hefei, China, July 2015.

actuators

Communication

Analytical Modeling and Simulation of S-Drive Piezoelectric Actuators

Nicholas A. Jones and Jason Clark *

Department of Electrical Engineering, Auburn University, Auburn, AL 36849, USA; alexj@ieee.org
* Correspondence: jvclark@auburn.edu

Abstract: This paper presents a structural geometry for increasing piezoelectric deformation, which is suitable for both micro- and macro-scale applications. New and versatile microstructure geometries for actuators can improve device performance, and piezoelectric designs benefit from a high-frequency response, power density, and efficiency, making them a viable choice for a variety of applications. Previous works have presented piezoelectric structures capable of this amplification, but few are well-suited to planar manufacturing. In addition to this manufacturing difficulty, a large number of designs cannot be chained into longer elements, preventing them from operating at the macro-scale. By optimizing for both modern manufacturing techniques and composability, this structure excels as an option for a variety of macro- and micro-applications. This paper presents an analytical compact model of a novel dual-bimorph piezoelectric structure, and shows that this compact model is within 2% of a computer-distributed element model. Furthermore it compares the actuator's theoretical performance to that of a modern actuator, showing that this actuator trades mechanical efficiency for compactness and weight savings.

Keywords: piezoelectric; actuator; robotic; MEMS

Citation: Jones, N.; Clark, J. Analytical Modeling and Simulation of S-Drive Piezoelectric Actuators. *Actuators* **2021**, *10*, 87. https://doi.org/10.3390/act10050087

Academic Editor: Shuxiang Dong

Received: 24 February 2021
Accepted: 23 April 2021
Published: 25 April 2021

1. Introduction

Applications of the inverse piezoelectric effect have been limited to a high-force, low-deflection regime due to the piezoelectric effect's generation of low strain, on the order of 0.1% [1] at a maximum, even for highly optimized materials. This low strain requires amplification of the deflection for a majority of applications. A number of designs exist for amplifying the deflection of a piezoelectric element [2], with the majority of these suffering from low active element density. These designs require a large amount of passive material to amplify to large deflections, decreasing the energy density of the overall actuator. In addition to their low density, these designs are rarely chained together to make longer actuators. This lack of flexibility and performance limits their usefulness in all but the most specific of applications.

One extant class of actuator which avoids these problems is telescoping actuators [3]. These actuators employ a series of meandering piezoelectric cylinders to gradually build displacement through the chain. Therefore, the displacement amplification is equal to the number of shells. While these actuators do excel in high-force and low-displacement applications, they are not suited to modern microfabrication techniques as they are highly non-planar. Planar designs operating on the same principles have been presented [4], but these still suffer from relatively low displacement amplification.

A second class of actuators that can greatly amplify generated strain, typically called flextensional actuators, does so through the trigonometric amplification of large contiguous piezoelectric elements [5–7]. This approach benefits from relatively simple designs, which are well-suited to planar manufacturing [8]. By varying geometric parameters, these actuators can be designed for a range of forces and deflections. The direction of actuation depends on the polarity of the d_{31} coefficient for the material used. This dependence means that a single material and geometry cannot be used to generate both expanding and contracting actuators that exceed the coercive fields [9]. Staying below the coercive field, while possible, reduces the maximum force for actuators substantially. Due to this limitation, actuators of this class needing to operate in the inverse of the material's piezoelectric coefficient suffer as compared to other actuator designs.

The final class of actuator discussed here is based on bimorphs. This class is both composable and amplifies strain to great extremes. Of the simple amplification schemes, bimorphs have the highest active to passive material ratio, making them ideal for large amplifying chains. Despite this advantage, simple bimorphs' curved displacement makes them unsuitable for chained geometries, necessitating modifications. By attaching two bimorphs end-to-end, the curvature can be counteracted [10,11]. These designs benefit from high geometric amplification ratios and compact chained element spacing. This paper presents and discusses an iteration on this type of actuator that is well-suited to modern planar manufacturing techniques.

2. Materials and Methods

As mentioned above, bimorph-based designs can easily improve on the basic geometry by joining two end-to-end. As can be seen in Figure 1, this modification creates a structure that, when actuated, forms a slight *S* shape, and more importantly, keeps both endpoints parallel throughout the range of travel. This parallelism facilitates the attachment of both loads or other s-drives to the structure. Typically, in a bimorph, the electrodes are separated along the width axis of Figure 1. This electrode arrangement is problematic, chaining multiple devices together, as it requires a complex routing of wires to connect subsequent elements. Thus, in this geometry, the electrodes are separated along the thickness axis instead of the width axis of the structure (Figure 1).

Notably, this geometry can use piezoelectric materials with both positive and negative piezoelectric coefficients simply by switching the elastic and piezoelectric layers. This flexibility allows for a manufacturing process to select any material and still make both contracting and expanding actuators. Additionally, the single poling direction for all piezoelectric regions allows for the poling to be done by the electrodes themselves, further simplifying the manufacturing process.

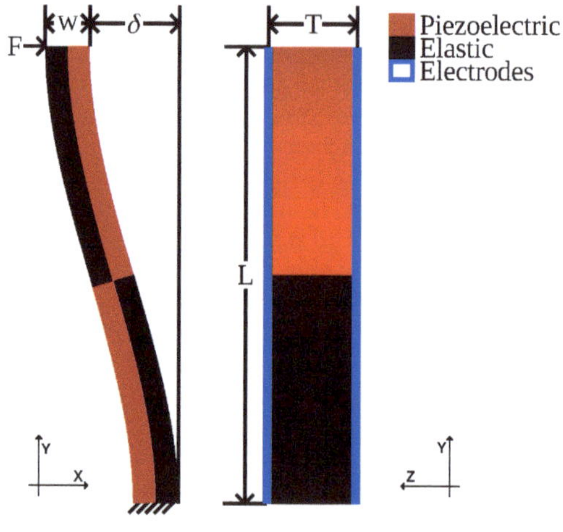

Figure 1. Geometry of a single actuated s-drive. F = load force, W = width, δ = deflection, T = thickness, and L = length. Electrodes omitted in XY view for clarity.

2.1. Analytical Model

Before s-drives can be used as micro- or macro-actuators, their constituent equations must be found. As with most electrical actuators, the primary terms of interest are blocking force and free deflection as a function of the input voltage.

2.1.1. Free Deflection

For deflection, an s-drive can be modeled as simply double that of one of its two corresponding heterogeneous bimorphs. Smits et al. [12] has (1). Modifying this relationship for an s-drive yields (2), which also compensates for the abnormal axis on which the electric field is applied in this geometry.

$$\delta = \frac{3Vd_{31}s_{11}^{p}s_{11}^{m}w_m(w_m+w_p)L^2}{K} \tag{1}$$

$$\delta_{\text{s-drive}} = \frac{3Vd_{31}s_{11}^{p}s_{11}^{m}(w_m^2w_p+w_p^2w_m)L^2}{2KT} \tag{2}$$

where K is:

$$K = 4s_{11}^{m}s_{11}^{p}\left(w_m(w_p)^3+w_p(w_m)^3+\frac{3(w_m)^2(w_p)^2}{2}\right)+(s_{11}^{m})^2(w_p)^4+(s_{11}^{p})^2(w_m)^4 \tag{3}$$

d_{31} is the piezoelectric coefficient. s_{11}^{p} and s_{11}^{m} are from the compliance matrices of the piezoelectric and elastic materials, respectively. w_p and w_m are the widths of the piezoelectric and elastic layer, respectively. For bulk elastic materials that have no compliance matrix, the young's moduli, E_p and E_m for piezoelectric and elastic, respectively, can be used to calculate compliance through

$$s_{11}^{m} = s_{11}^{p}\frac{E_p}{E_m} \tag{4}$$

2.1.2. Blocking Force

While Smits et al. [12] have equations for the blocking force of a bimorph, those assume that rotation of the tip is unconstrained. This assumption does not hold true for an s-drive. Instead, Euler–Bernoulli beam theory can be used to find the stiffness of a cantilever as

$$k_{\text{cantilever}} = \frac{3EI}{L^3} \tag{5}$$

where E is the Young's modulus and I is the second moment of area for the actuator. A full s-drive does not take the same shape as a cantilever but each half does, and so the full s-drive can be approximately modeled as two cantilevers of length $\frac{L}{2}$ in series, yielding

$$k_{\text{s-drive}} = \frac{3EI}{2\left(\frac{L}{2}\right)^3} = \frac{12EI}{L^3} \tag{6}$$

Furthermore, EI in (6) must account for the differing Young's moduli in the two materials. This adjustment involves calculating the second moment of area of the two material halves about the central axis of the beam, yielding

$$EI = \frac{E_mT((w_m)^3+3w_m(w_p)^2)}{12} + \frac{E_pT((w_p)^3+3w_p(w_m)^2)}{12} \tag{7}$$

With the stiffness and deflection calculated, blocking force is the product of (6) and (2).

$$F_{\text{blocking}} = k\delta = \frac{18V(EI)d_{31}s_{11}^{p}s_{11}^{m}w_m w_p(w_m + w_p)}{KTL} \tag{8}$$

2.1.3. Chain Stiffness

In certain MEMS applications, and certainly in macro-applications, chaining multiple actuators in sequence may be required to achieve the desired deflection. Figure 2 shows a possible geometry for realizing this chain. Figure 3 shows this geometry both being chained in series and working in parallel. Notably, due to the abnormal axis of the applied electric field, this design does not need additional wires between chained elements, as the electrodes are contiguous, simplifying manufacturing.

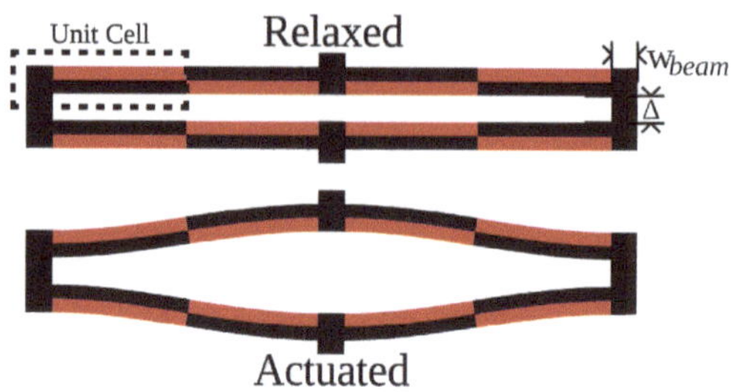

Figure 2. Unactuated and actuated composable s-drives.

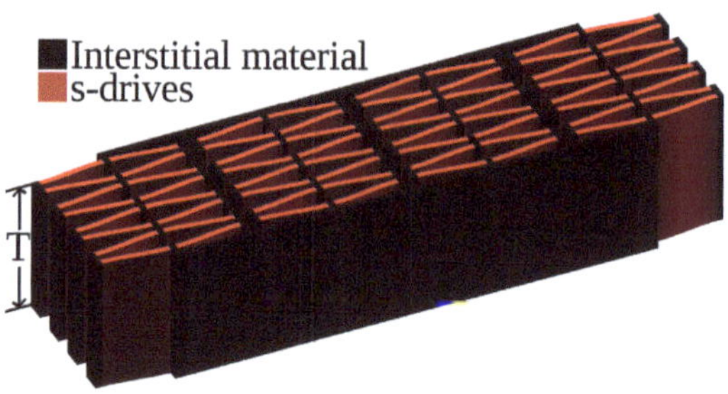

Figure 3. S-Drives when used in series and parallel.

The free deflection of such a geometry is $N\delta$, where N is the number of s-drives in series. Blocking force is not as trivial, as the connecting beams add compliance to the structure, reducing the total stiffness. Balakrisnan et al. [13] modeled meandering springs, and the approach found the stiffness of a single unit cell, which is defined in Figure 2. The bulk stiffness is the combination of all the unit cells, and the unit cell stiffness is the cantilever stiffness combined with that of the adjoining beam connected to a rigid moment arm. A more thorough derivation can be found in Balakrisnan et al., but the final form for this stiffness is in (9). Combining (9) with that of the cantilever (5) is (10) and the entire geometry is (11):

$$k_{\text{adjoining}} = \frac{2E_m T(w_{\text{beam}})^3}{3\left(\frac{\Delta}{2} + w_p + w_m\right)(L + w_{\text{beam}})^2} \tag{9}$$

$$k_{\text{unit}} = \frac{k_{\text{s-drive}}k_{\text{adjoining}}}{k_{\text{s-drive}} + k_{\text{adjoining}}} \tag{10}$$

$$k_{\text{total}} = \frac{k_{\text{unit}}}{2} \tag{11}$$

This model from Balakrisnan et al., only works well when $E_m \approx E_p$ and $W_{\text{beam}} \approx W$.

3. Results

3.1. Analytical Model Versus Simulations

COMSOL Multiphysics® was used to verify the model. The geometry in Figure 2 was imported into the software, with a variety of geometries that can be seen in Table 1. The material properties applied were the built-in specifications for PZT-5H and aluminum for the piezoelectric and elastic layers, respectively. The length parameter for each geometry was varied from 50 µm to 400 µm, while the thickness and applied voltage were held constant at 2 µm and 8 V, respectively. This method generates an electric field of 4 MV m^{-1} which is approximately the upper limit of strain generation in PZT-5H, about 0.1% [14].

Comparing the results from COMSOL® with the model developed, the error is typically well below 2%, as can be seen in Figure 4.

Table 1. Dimensions of geometries tested

Geometry	A	B	C
w_m	2 µm	2 µm	1 µm
w_p	2 µm	2 µm	3 µm
Δ	3 µm	10 µm	2 µm

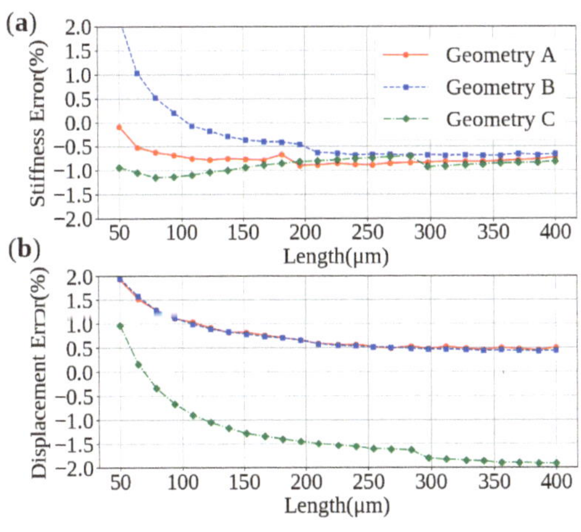

Figure 4. Percent error for: (**a**) stiffness and (**b**) deflection.

3.2. Performance Versus a Modern Actuator

As a point of comparison with other works, an S-Drive was designed with the same piezoelectric mass (6.6 mg) as that of York et al. [6]. The materials used in simulator were also set to be those reported in York et al., (PZT-5H and Alumina) and the same electric field of 1.5 V/µm was applied. As can be seen in Table 2, the design presented in York et al., is stiffer but with lower mass efficiency and less compactness along the axis of actuation. This difference results in the S-Drive being better suited for regimes where actuation authority and mass are more important than total output force.

Further included in Table 2 are the performance data for our group's prior work [15]. This new geometry shows marked improvements in every category, including more than double the blocking force.

Table 2. Performance factors comparison for equivalent active elements.

	This Work	Our Prior Work	York et al.
Free Displacement (μm)	88.9	89.4	89.6
Blocking Force (mN)	130	60	211
Mass (mg)	8.3	8.5	16.8
Thickness (μm)	270	270	270
Width (mm)	11.3	12.6	≥8
Length (mm)	0.8	0.83	≥2
Effective Strain (%)	11.1	10.7	4.1

4. Discussion

Advances in manufacturing have made this design substantially more feasible to produce compared to its introduction in Ervin et al. [10]. This design's need for fine-grained mixing of both piezoelectric and non-piezoelectric material rules out a large number of processes, both old and new.

One potentially viable manufacturing approach is to micromachine layers of the different materials in the correct geometry, and then assemble the different layers, as in York et al. [6]. For this use case, this approach has the substantial drawback of requiring adhesive between layers to provide sufficient coupling between the elastic and piezoelectric layers. The adhesive would act as a parasitic element and the microassembly would hamper the automation of production.

Alternatively, the use of a single material would avoid the substantial drawbacks of both adhesive and assembly. By patterning the electrodes only over the active regions of the geometry, the actuator can be realized from a single bulk piece of piezoelectric material. Using a single material would drastically simplify both the model and the manufacturing. The performance of the actuator would suffer slightly from stray fields bleeding into the inactive region, but that could be compensated for by reducing the width of the electrodes. This approach is the most promising for mass production due to its simplicity.

In circumstances where micromachining is not accessible, 3D printing may be used in its place. This actuator's performance depends greatly on the elastic modulus of its constituent materials. Commonly accessible polymers are highly compliant. To rectify this, carbon fiber can be added to the polymers to drastically improve stiffness, with 13% carbon fiber yielding a 400% increase in elastic modulus in Love et al. [16]. Additionally, the polymers can be made conductive, as in Leigh et al. [17], and piezoelectric, as in Cholleti [18], allowing for all steps of the manufacturing process to be completed on a single multi-material capable printer. Optimizing the materials for maximum possible stiffness is key to producing actuators of any merit. Once optimized, these high stiffness materials would allow for home manufacturing of actuators with only minimal equipment costs.

5. Conclusions

This paper has presented a piezoelectric actuator geometry capable of large deflection along with a mathematical model for its performance. While well-suited to micro-applications, its compactness, weight savings, and composability permit it to scale into macro-applications as well. With the use of modern manufacturing, this design is well-suited to mass production, while further optimization of material and technique would allow for manufacture with more affordable equipment.

Author Contributions: Writing, N.A.J.; review and editing, N.A.J. and J.C.; visualization, N.A.J., supervision, J.C.; project administration, J.C.; conceptualization, N.A.J. and J.C.; methodology, N.A.J.; verification, N.A.J., J.C.; formal analysis, N.A.J.; investigation, N.A.J. and J.C.; original draft preparation, N.A.J. All authors have read and agreed to the published version of the manuscript.

Funding: This research received no external funding.

Institutional Review Board Statement: Not applicable.

Informed Consent Statement: Not applicable.

Data Availability Statement: No new data were created or analyzed in this study. Data sharing is not applicable to this article.

Conflicts of Interest: The authors declare no conflict of interest.

References

1. Huang, Y.; Zhang, S.; Wang, P.; Xia, Y.X.; Lin, D.H.; Yao, K.; Lim, L.-C. Hi-Fi Stake Piezo Single Crystal Actuator. *Actuators* **2018**, *7*, 60. [CrossRef]
2. Wang, S.; Rong, W.; Wang, L.; Xie, H.; Sun, L.; Mills, J.K. A survey of piezoelectric actuators with long working stroke in recent years: Classifications, principles, connections and distinctions. *Mech. Syst. Signal Process.* **2019**, *123*, 591–605. [CrossRef]
3. Wu, C.; Lewis, D.; Kahn, M.; Chase, M. High-authority telescoping actuators. In Proceedings of the SPIE 3674, 1999 Symposium on Smart Structures and Materials, Newport Beach, CA, USA, 1–4 March 1999.
4. Robbins, W.P.; Polla, D.L.; Glumac, D.E. High-displacement piezoelectric actuator utilizing a meander-line geometry I. Experimental characterization. *IEEE Trans. Ultrason. Ferroelectr. Freq. Control* **1991**, *38*, 454–460. [CrossRef] [PubMed]
5. Dogan, A.; Uchino, K.; Newnham, R.E. Composite piezoelectric transducer with truncated conical endcaps "cymbal". *IEEE Trans. Ultrason. Ferroelectr. Freq. Control* **1997**, *44*, 597–605. [CrossRef]
6. York, P.; Jafferis, N.; Wood, R. Millimeter-sized piezoelectric flextensional actuators with improved mechanical efficiency. *Sens. Actuators A Phys.* **2020**, *311*, 112066. [CrossRef]
7. Sun, F.; Hao, Y.; Xu, F.; Jin, J.; Li, Q.; Tong, L.; Zhang, M.; Zhang, X. Proposal of an Equal-Stiffness and Equal-Stroke 2D Micro-Positioning Platform Driven by Piezoelectric Actuators. *Actuators* **2020**, *9*, 47. [CrossRef]
8. Toledo, J.; Ruiz-Díez, V.; Diaz-Molina, A.; Ruiz, D.; Donoso, A.; Bellido, J.C.; Wistrela, E.; Kucera, M.; Schmid, U.; Hernando-García, J.; et al. Design and Characterization of In-Plane Piezoelectric Microactuators. *Actuators* **2017**, *6*, 19. [CrossRef]
9. Secord, T.W.; Ueda, J.; Asada, H.H. Dynamic analysis of a high-bandwidth, large-strain, PZT cellular muscle actuator with layered strain amplification. In Proceedings of the 2008 IEEE International Conference on Robotics and Automation, Pasadena, CA, USA, 19–23 May 2008; pp. 761–766.
10. Ervin, J.D.; Brei, D. Recurve piezoelectric-strain-amplifying actuator architecture. *IEEE/ASME Trans. Mechatron.* **1998**, *3*, 293–301. [CrossRef]
11. Beckers, G.; Pirnay, L.; Dehez, B. Design and modelling of a ring-shape Recurve piezoelectric actuator. In Proceedings of the ISAF-ECAPD-PFM, Aveiro, Portugal, 9–13 July 2012; pp. 1–4.
12. Smits, J.G.; Choi, W. The constituent equations of piezoelectric heterogeneous bimorphs. *IEEE Trans. Ultrason. Ferroelectr. Freq. Control* **1991**, *38*, 256–270. [CrossRef] [PubMed]
13. Balakrisnan, B.; Nacev, A.; Burke, J.M.; Dasgupta, A.; Smela, E. Design of Compliant Meanders for Applications in MEMS, Actuators, and Flexible Electronics. *Smart Mater. Struct.* **2012**, *21*, 075033. [CrossRef]
14. Giurgiutiu, V.; Pomirleanu, R.; Rogers, C. *Energy-Based Comparison of Solid-State Actuators*; Laboratory for Adaptive Materials and Smart Structures, University of South Carolina: Columbia, SC, USA, 2000.
15. Chigullapalli, A. Modeling and Validation of S-Drive: A Nestable Piezoelectric Actuator (2014). Open Access Dissertations. 1501. Available online: https://docs.lib.purdue.edu/open_access_dissertations/1501 (accessed on 12 March 2020).
16. Love, L.J.; Kunc, V.; Rios, O.; Duty, C.E.; Elliott, A.M.; Post, B.K.; Smith, R.J.; Blue, C.A. The importance of carbon fiber to polymer additive manufacturing. *J. Mater. Res.* **2014**, *29*, 1893–1898. [CrossRef]
17. Leigh, S.J.; Bradley, R.J.; Purssell, C.P.; Billson, D.R.; Hutchins, D.A. A Simple, Low-Cost Conductive Composite Material for 3D Printing of Electronic Sensors. *PLoS ONE* **2012**, *7*, e49365. [CrossRef]
18. Cholleti, E. A Review on 3D printing of piezoelectric materials. *IOP Conf. Mater. Sci. Eng.* **2018**, *455*, 012046. [CrossRef]

actuators

MDPI

Article

Sense–Analyze–Respond–Actuate (SARA) Paradigm: Proof of Concept System Spanning Nanoscale and Macroscale Actuation for Detection of *Escherichia coli* in Aqueous Media

Cassie A. Giacobassi [1], Daniela A. Oliveira [1], Cicero C. Pola [2], Dong Xiang [3,4], Yifan Tang [3], Shoumen Palit Austin Datta [3,5,6,7], Eric S. McLamore [3,*] and Carmen L. Gomes [2,*]

1 Department of Biological and Agricultural Engineering, Texas A&M University, College Station, TX 77843, USA; c.giacobassi@gmail.com (C.A.G.); daoliveira@tamu.edu (D.A.O.)
2 Department of Mechanical Engineering, Iowa State University, Ames, IA 50011, USA; cicerocp@iastate.edu
3 Agricultural and Biological Engineering, Institute of Food and Agricultural Sciences, University of Florida, Gainesville, FL 32611, USA; xiangdong@ufl.edu (D.X.); tang.yifan@ufl.edu (Y.T.); shoumen@mit.edu (S.P.A.D.)
4 Centre for Life Sciences, School of Integrative Sciences and Engineering, National University Singapore, Singapore 119077, Singapore
5 MIT Auto-ID Labs, Department of Mechanical Engineering, Massachusetts Institute of Technology, Cambridge, MA 02139, USA
6 MDPnP Labs, Biomedical Engineering Program, Department of Anesthesiology, Massachusetts General Hospital, Harvard Medical School, Cambridge, MA 02139, USA
7 NSF Center for Robots and Sensors for Human Well-Being (RoSeHuB), Collaborative Robotics Lab, 193 Knoy Hall, Purdue University, West Lafayette, IN 47907, USA
* Correspondence: emclamore@ufl.edu (E.S.M.); carmen@iastate.edu (C.L.G.); Tel.: +1-352-294-6703 (E.S.M.); +1-515-294-1138 (C.L.G.)

Abstract: Foodborne pathogens are a major concern for public health. We demonstrate for the first time a partially automated sensing system for rapid (~17 min), label-free impedimetric detection of *Escherichia coli* spp. in food samples (vegetable broth) and hydroponic media (aeroponic lettuce system) based on temperature-responsive poly(N-isopropylacrylamide) (PNIPAAm) nanobrushes. This proof of concept (PoC) for the Sense-Analyze-Respond-Actuate (SARA) paradigm uses a biomimetic nanostructure that is analyzed and actuated with a smartphone. The bio-inspired soft material and sensing mechanism is inspired by binary symbiotic systems found in nature, where low concentrations of bacteria are captured from complex matrices by brush actuation driven by concentration gradients at the tissue surface. To mimic this natural actuation system, carbon-metal nanohybrid sensors were fabricated as the transducer layer, and coated with PNIPAAm nanobrushes. The most effective coating and actuation protocol for *E. coli* detection at various temperatures above/below the critical solution temperature of PNIPAAm was determined using a series of electrochemical experiments. After analyzing nanobrush actuation in stagnant media, we developed a flow through system using a series of pumps that are triggered by electrochemical events at the surface of the biosensor. SARA PoC may be viewed as a cyber-physical system that actuates nanomaterials using smartphone-based electroanalytical testing of samples. This study demonstrates thermal actuation of polymer nanobrushes to detect (sense) bacteria using a cyber-physical systems (CPS) approach. This PoC may catalyze the development of smart sensors capable of actuation at the nanoscale (stimulus-response polymer) and macroscale (non-microfluidic pumping).

Keywords: *Escherichia coli*; lectin; thermo-responsive polymer; food safety; biosensor; artificial reasoning tools (ART); sensor-analytics point solutions (SNAPS); Sense-Analyze-Respond-Actuate (SARA); percepts-environment-actuators-sensors (PEAS); cyber-physical systems (CPS)

Citation: Giacobassi, C.A.; Oliveira, D.A.; Pola, C.C.; Xiang, D.; Tang, Y.; Datta, S.P.A.; McLamore, E.S.; Gomes, C.L. Sense–Analyze–Respond–Actuate (SARA) Paradigm: Proof of Concept System Spanning Nanoscale and Macroscale Actuation for Detection of *Escherichia coli* in Aqueous Media. *Actuators* **2021**, *10*, 2. https://dx.doi.org/10.3390/act10010002

Received: 28 October 2020
Accepted: 19 December 2020
Published: 23 December 2020

1. Introduction

Foodborne pathogens in the global food supply chain increase the risk of mortality and morbidity. Reducing the resulting economic and public health burden calls for new technologies to prevent disease outbreaks; see critical reviews [1–4]. According to the National Outbreak Reporting System (NORS), an annual average of 15,332 illnesses, 879 hospitalizations, and 22 deaths caused by foodborne diseases were reported from 2008 to 2017 [5]. Additionally, the U.S. Centers of Disease Control and Prevention (CDC) [6] estimates that around 48 million people get sick, 128,000 are hospitalized, and 3000 deaths occur in the U.S. each year from foodborne diseases. Such discrepancy emphasizes the limitations of data collection correlated with misinformation, misreport, and lack of accessible diagnostic tools. Among the food products commonly associated with outbreaks, fresh produce is one of the leading causes of foodborne illnesses [7], with 377 outbreaks reported by the CDC from 2004 to 2012 [8]. The bacteria commonly associated with foodborne outbreaks include *Salmonella* spp., *Listeria monocytogenes*, and pathogenic *Escherichia coli.*

The presence of *E. coli* in water, food, and food contact surfaces is used as evidence for fecal contamination. Among the hundreds of *E. coli* strains, pathogens are commonly classified as either verotoxigenic *E. coli* (VTEC) or Shiga-toxin-producing *E. coli* (STEC) [9]. Toxins produced by VTEC and STEC are known to cause damage to the intestinal lining, disrupts the homeostasis of the gastrointestinal tract microbiota, and are responsible for symptoms such as hemorrhagic colitis, renal failure, and hemolytic anemia. The most common pathogenic strain is *E. coli* O157:H7, which has an infectious dose of 10-100 cells [10,11]. The U.S. Food and Drug Administration (FDA) identified six serogroups of pathogenic *E. coli* commonly referred to as the "big six": *E. coli* O26, O45, O103, O111, O121, and O145. Each of these serogroups differs in terms of the O-antigen surface structure, which is a critical lipopolysaccharide found on the outer membrane of *Gram*-negative bacteria [12].

Recent multistate outbreaks of *E. coli* O157:H7 infections linked to romaine lettuce demonstrate the need for improved monitoring methods and management strategies [13]. The Food Safety Modernization Act (FSMA) addressed this issue with a series of directives released between 2015 and 2020. In general, the directives were intended to change the U.S. food systems focus from a reactive or responsive safety paradigm to a preventative safety paradigm [14–17]. The role of sensors/detectors within the modern food supply chain has been reviewed by several groups [18–21].

For irrigation water, the *E. coli* concentration (both pathogenic strains and non-pathogenic strains) must not exceed 126 CFU/100 mL (geometric mean) without triggering a responsive action [22]. Several recent reviews have addressed the technological challenges of implementing these federal guidelines for the specific case of alternative water sources (e.g., treated wastewater, brackish water) [23,24]. For postharvest, in general, there is a requirement of no detectable generic *E. coli* in 100 mL of water used on food contact surfaces or in direct contact with produce [22]. The implementation of a "hold and test" policy in 2012 by the U.S. Department of Agriculture (USDA) significantly reduced the risk of consumer exposure to unsafe products via food recalls [25]. However, food recalls are onerous for food companies, cause significant economic loss [26], and exacerbate food waste [27]. Keener et al. [28] promoted food safety paradigms that blend regulation and legislation, a notion that is only possible if data-informed decision support tools are actively used [19,29].

The current standard methods used by the food industry include culture and colony counting, polymerase chain reaction (PCR), and enzyme-linked immunosorbent assay (ELISA) to detect foodborne pathogens [2,3]. These diagnostic tools require several hours to days to provide a result, which exacerbates problems associated with the current test and hold policies. As food safety regulations become stricter, there is a pressing need for improved tools to ensure food safety without asphyxiating the economic needs of the food industry. Rapid (<1 h) and accurate detection methods for foodborne pathogens that provide high-throughput screening capability and inform decision support systems are pivotal across the food supply chain, from production to processing and distribution (i.e., from "farm to fork") [30–32]. To meet this need, a plethora of different biosensors have been developed. Tools using the CRISPR system [33] are expected in the future.

Pathogen Biosensors in Food Systems

Biosensors to detect bacteria in the food supply chain are ideal if designed to serve as rapid, portable devices that can be used for screening food products, either targeting unique extracellular structures or proteins, peptides, glycans, and genetic material. The most common approaches for biosensors targeting extracellular structures use either fluorescent, colorimetric, or electrochemical sensing modalities. Comprehensive reviews on the topic summarize recent developments in culture-independent techniques [3], commercially available sensor systems [1], and emerging materials for biosensor development [2]. Electrochemical biosensors are generally unaffected by optically dense fluids, and the transduced signal is directly correlated with analyte concentration without an additional signal conversion technology such as a photodiode. These features are important when considering the analysis of food samples where simple hardware systems and detection protocols are necessary [19,34–37]. To improve the performance of biosensors in complex food matrices, biorecognition agents (e.g., antibodies, aptamers, lectins, peptides) are commonly immobilized on transducer nanomaterials such as nanometals (palladium, platinum, gold, nickel, copper) and nanocarbon (graphene, carbon nanotubes, carbon nanodots) [38–42]. Among these hybrid nanomaterials, metallic-nanocarbon composites have demonstrated improved electron transport and stability, consequently enhancing sensitivity, response time, and limit of detection [43–48].

Hills et al. [49] recently developed a pathogen biosensor using stimulus-response polymer nanobrushes conjugated to a platinum-nanocarbon electrode. The technique was used to measure *Listeria* using a DNA aptamer conjugated to the terminal group on a chitosan nanobrush. The work showed that by controlling nanobrush actuation at the microscale bacteria can be captured with the brush in the extended state, and subsequent collapse of the nanobrush improves signal transduction (actuation was controlled via solution pH). The ability to induce structural changes on the biosensor surface facilitates dynamic control over the Debye layer and therefore signal transduction. The major advantage relative to the use of passive sensor materials is an increase in the probability of target–receptor interactions due to micron scale structural changes. To date, this approach has not been demonstrated in samples larger than 20 mL, which is critically important for application in food production systems. The use of chitosan nanobrush sensors may not be viable for large sample volumes due to the need for large volumes of acid/base to modulate pH.

Poly(N-isopropylacrylamide) (PNIPAAm) is a stimulus-response polymer that expands at temperatures below the lower critical solution temperature (LCST) and collapses at temperatures above the LCST, which is around 32–35 °C [50]. This behavior, which is observed above and below the LCST, is fully reversible in aqueous solution due to the changes in the hydrogen-bonding interfaces of the amide group [51]. PNIPAAm is also a beneficial polymer for biosensing as different terminal groups can be added (e.g., carboxyl or amine), facilitating the conjugation of a receptor [52]. Previous works with PNIPAAm stimulus-response behavior focused on fundamental studies of bioreceptor loading [41] or electronic behavior [53].

This study demonstrates for the first time the development of partially automated system for rapid (~17 min), label-free impedimetric biosensor for real-time detection of *E. coli* spp. in food samples (i.e., vegetable broth) and hydroponic media (aeroponic lettuce system) based on stimulus-response of PNIPAAm nanobrushes. Additionally, the semi-automated system demonstrated herein allows for continuous detection of *E. coli* and potentially other foodborne pathogens in aqueous media reducing sample handling problems, and consequently helping mitigating disease outbreaks. Two well-known bioreceptors, namely antibodies and lectins, were tested as bioreceptors. The biomaterial was tested on two different metal-nanocarbon hybrid electrodes. Furthermore, we show that actuation of PNIPAAm nanobrushes leads to enhanced bacteria capture and controllable electrochemical transduction based on an external stimulus (i.e., temperature) in laboratory studies. Finally, we demonstrate a proof of concept Sense-Analyze-Respond-Actuate (SARA) system based on actuation and analysis using a remote device (smartphone) (see supplemental materials Figure S1).

2. Materials and Methods

2.1. Materials and Bacteria Cultures

Lead acetate, chloroplatinic acid, 2-aminoethanethiol hydrochloride (AESH), ascorbic acid, potassium phosphate monobasic, Concanavalin A (ConA) from *Canavalia ensiformis* (Jack bean), sodium chloride, sodium phosphate dibasic, 11-mercaptoundecanoic acid (11-MUA), Anti-GroEL antibody (Ab) produced in rabbit, and potassium chloride were purchased from Sigma-Aldrich (St. Louis, MO, USA). Calcium chloride, sodium persulfate ($Na_2S_2O_8$), sodium nitrate ($NaNO_3$), 1-ethyl-3-(3-dimethylaminopropyl) carbodiimide HCl (EDC), and manganese chloride were obtained from Thermo Fisher Scientific (Waltham, MA, USA). Sulfuric acid (H_2SO_4) was purchased from Fisher Scientific (Potassium ferrocyanide trihydrate was purchased from Ward's Science (Rochester, NY, USA). N-Hydroxysuccinimide (NHS), potassium nitrate, 2-(morpholino) ethanesulfonic acid (MES) buffer, glutaraldehyde, and platinum wire (99.95% Pt, 0.5 mm dia.) were obtained from Alfa Aesar (Ward Hill, MA, USA). Single-layered graphene oxide (GO) was purchased from ACS Material (Medford, MA, USA). N-Isopropylacrylamide (NIPAAm) was obtained from Tokyo Chemical Industry Co. (Portland, OR, USA). Buffered peptone water (BPW) and tryptic soy broth (TSB) were acquired from Becton, Dickson and Company (Sparks, MD, USA). Petrifilm-Aerobic Count Plates were purchased from 3M (St. Paul, MN, USA). Platinum/iridium (Pt/Ir) electrodes, reference electrodes (Ag/AgCl) and Pt auxiliary electrodes were purchased from BASi, Inc. (West Lafayette, IN, USA). Gold interdigitated electrodes (3 mm × 5 mm; 100 μm gap spacing) were purchased from DropSens (Asturias, Spain). Screen printed carbon electrodes (SPC, 5 mm diameter) were purchased from Zensor (model SE100, Zensor USA, Katy, TX, USA).

Sensitivity and selectivity were tested using *Escherichia coli* (ATCC 35218, Manassas, VA, USA) in phosphate buffer saline (PBS, pH 7.4). For selectivity testing, *Salmonella enterica* serovar Enteritidis (ATCC BAA-1045, Manassas, VA, USA) was used in PBS. *Escherichia coli* O157:H7 (ATCC 43895, Manassas, VA, USA) was used for testing in vegetable broth and hydroponic systems. All bacteria initially stored at −80 °C were replicated by identical duplicate transfers and incubated under aerobic conditions for 24 h at 35 °C. The bacterial cultures were maintained on TSA (tryptic soy agar) slants at 4 °C. Transfers from slants were conducted to prepare microorganisms for analysis. Bacteria samples were serially diluted in BPW and enumerated on Petrifilm™ aerobic count plates (3M, Saint Paul, MN, USA) after 48 h at 35 °C; results are reported as CFU/mL.

2.2. Image Analysis

Electrode morphology was imaged using field emission scanning electron microscopy (SEM) with a FEI Quanta 600 FEG (Hillsboro, OR, USA). All electrodes were first coated with a 10 nm thick layer of platinum using a Cressington sputter coater 208 HR (Watford, United Kingdom). Electrodes were retrieved from the sputter coater and allowed to ventilate for 30 min prior to imaging. Images were obtained at magnifications of 5000× and 10,000× and 5 kV. Field SEM images were used to determine average particle size based on post-measurement analysis with ImageJ.

Scanning white light interferometry (SWLI) was used to determine the average roughness profile of each material using an Alicona Infinite Focus Microscope (G4 Optical 3D Surface Profiler). A 20× objective with a focus-variation system and a lateral resolution of 0.88 μm was used for all scans (Bartlett, IL). SWLI was used to determine profile average roughness as well as contour topology.

2.3. Nanomaterial Deposition and Electrode Biofunctionalization

Prior to use, all Pt/Ir electrodes were polished and cleaned according to previously reported methods [41]. Pt/Ir electrodes were modified through the application of a graphene-platinum nanocomposite layer following procedures adapted from Vanegas et al. [42] and Hills et al. [49]. Briefly, a first layer of nanoplatinum (nPt) was formed on the surface of each electrode via sonoelectrodeposition at 10 V for 90 s in a solution of 1.44% (w/v) chloroplatinic acid and 0.002% (w/v) lead acetate. Next, a reduced graphene oxide (rGO) layer (2 mg/mL in DI) was drop coated onto the surface of the nPt-modified electrode and dried for 30 s at 40 °C with an 1875 W heat gun (Revlon, New York, NY). The semi-dried electrode was then spun in the spin coater for 30 s at 1700 rpm and then for 60 s at 3500 rpm. Lastly, a second layer of nPt was electrodeposited onto the surface to complete the "sandwich" structure using the same methods as above. The SPC electrodes used for the SARA proof of concept in hydroponic system were also initially cleaned using cyclic voltammetry (see Supplementary Materials).

PNIPAAm nanobrushes were deposited on both electrodes (Pt/Ir and SPC) based on Zhao et al. [54]; in our study a potentiostat from CH Instruments (Model 600 E Series) was used throughout. Briefly, 1 M NIPAAm, 0.2 M $NaNO_3$, 0.01 M $Na_2S_2O_8$, and 4.85 mM AESH were suspended in 20 mL RO water to prepare the NIPAAm solution. The AESH serves as a chain transfer agent to have the amine end group required for subsequent attachment of bioreceptors, namely lectin (ConA) or antibody (Ab) [52]. Polymerization of PNIPAAm-NH_2 onto the electrode was achieved at room temperature using cyclic voltammetry (CV) with the following settings: potential range from −0.35 V to −1.35 V and a scan rate of 100 mV/s for 60 cycles [54].

To conjugate ConA or anti-GroEL antibody to PNIPAAm nanobrushes, nPt-rGO-PNIPAAm-modified electrodes were incubated with an aqueous solution containing glutaraldehyde in a 2:1 molar ratio to AESH and allowed to react under agitation for 2 h at room temperature (resulting in amine-amine bonds). Electrodes were then exposed to ConA or anti-GroEL antibody suspensions at either 50 nM, 100 nM, or 200 nM. Ca^{2+} and Mn^{2+} ions were added to the PBS solution to promote carbohydrate binding and achieve optimum ConA activity [55,56]. After all conjugation steps, the unbound recognition agents were washed off with 100 µL of 1× PBS thrice. For the attachment of ConA and anti-GroEL antibodies to nPt-rGO-modified electrodes, the surface was initially carboxylated with a self-assembled monolayer (SAM) using 11-MUA, followed by amine-carboxyl conjugation chemistry (see Supplementary Materials). All modified electrodes were stored in PBS (pH 7.4) at 4 °C until further analysis. The conjugation of ConA to the SPC-PNIPAAm nanobrushes electrodes was performed similarly. After 2 h incubation with the glutaraldehyde/AESH solution, ConA suspension, in the presence of Ca^{2+} and Mn^{2+}, was applied to the SPC electrodes for 2 h under agitation. Then, functionalized SPC electrodes were freeze-dried (see Supplementary Materials) and stored at −20 °C until use for testing in the hydroponic system.

2.4. Electrochemical Analysis

Laboratory electrochemical characterization was performed using a three electrodes cell stand according to previous work [42,49,57]. Cyclic voltammetry (CV) was carried out in 4 mM $Fe(CN)_6{}^{3-}/Fe(CN)_6{}^{4-}$ (1:1) redox probe with 1 mM KNO_3 solution at a switching potential of 650 mV versus a Ag/AgCl reference electrode with 30 s quiet time at scan rates of 50, 100, 150, and 200 mV/s. The electroactive surface area (ESA) was determined using the Randles-Sevcik equation by plotting the current (I) versus the square root of scan rate ($v^{1/2}$), as described previously [40,42,58].

Electrochemical impedance spectroscopy (EIS) tests were conducted in a solution of 4 mM $Fe(CN)_6{}^{3-}/Fe(CN)_6{}^{4-}$ in 1 M KCl with an AC amplitude of 0.1 V. An initial DC potential of 0.25 V was applied with a frequency range of 1–100,000 Hz [41,49,59]. Complex plane diagrams (Nyquist plots) were used to determine the charge transfer resistance (R_{ct}) based on equivalent circuit analysis (Randles-Sevcik circuit).

PNIPAAm nanobrush actuation was evaluated through CV and EIS. These tests were carried out at temperatures above and below PNIPAAm's LCST, namely 40 °C and 20 °C, respectively. A water bath was used to control the temperature throughout the measurements. The PNIPAAm nanobrushes' response to change in stimuli was used to determine the most efficient conditions for bacteria capture and sensing was based on the protocol for brush actuation by Hills et al. [49], which was based on analyzing ESA and R_{ct} results for each nanobrush-receptor material.

2.5. Biosensor Performance Testing

EIS was used to determine the limit of detection (LOD), range, sensitivity, and selectivity of each biosensor when exposed to bacteria at concentrations varying from 10–10^8 CFU/mL. Bode plots were used to determine the impedance at a fixed cutoff frequency, which is 1 Hz. Where noted, the change in impedance ($\Delta Z = Z_{bacteria} - Z_{no\ bacteria}$) was determined from the Bode plots. Sensitivity to the target bacterium was determined by the slope of the linear portion of the calibration curve consisting of the change in impedance (Ω) vs. the concentration of cells (log CFU/mL) [49,60]. Sensitivity testing with the optimized ConA and antibody sensors was performed in PBS, sterile vegetable broth (purchased from a local market), and hydroponic media. Selectivity to *E. coli* O157:H7 was measured by determining sensitivity, LOD, and range in the presence of *Gram*-negative bacteria in PBS (*E. coli* 35218, *E. coli* 43895, and *Salmonella* Typhimurium) [60]. The LOD was calculated using the 3σ method, and the range was calculated as the linear portion of the calibration curve ($R^2 > 0.98$).

2.6. SARA Paradigm Proof of Concept in Hydroponic System

SPC biosensors with a handheld potentiostat [37] were used to construct a proof of concept SARA system for hydroponic lettuce. Hydroponic sensing experiments were performed using a similar setup as described by Sidhu et al. [61]. A RainForest modular 318 aeroponic system with Vortex sprayer was used to grow lettuce based on Marhaenanto et al. [62]. The main reservoir of the hydroponic system was 50 L and the conical vortex sprayer was operated at 1200 rpm. Hydroponic lettuce (*Lactuca saliva*) was cultivated using 7.6-cm-diameter plastic seed cups with CocoTek liners and expanded clay pellets (Mr. Stacky Hydroponic Center, Lake City, FL, USA). Full spectrum LED grow lights (75 W equivalent) were used (photoperiod of 8 h). Nutrient solution (Liquid Plant Food Big Bloom, Fox Farm Organic Gardening, Arcata, CA, USA) was replaced every 7 days based on manufacturer's recommendations. Growth media was sterilized according to our previous methods [63]

A particle trap was spliced into a $\frac{3}{4}$" OD Tygon tube and attached to a submersible pump for biosensor measurements. The particle trap had a stainless-steel mesh (#50; 300 μm mesh) within the inner chamber, and the biosensor was fixed within this mesh strain for direct contact with the water prior to filtration in the particle trap. The trap was customized for biosensor analysis by drilling two small holes on the top of the plastic housing and threading male-male Dupont Wire (Arduino) through the hole. The holes in the plastic body were sealed with rubber sealant (FlexSeal, Weston, FL, USA) and the inner pins were soldered to the sensor bonding pads. Lastly, the lead wires were insulated with nail polish and dried overnight, and then the trap was fixed to the housing and sealed via the threaded fitting.

A schematic diagram of the hydroponic system with the SARA sampling loop is shown in supplemental Figure S3. The hydroponic system contained a sampling loop with the trap, and a micropump was used to manually extract samples from the main 50-L reservoir at a rate of 10 mL/min. The sampling loop for SARA included a micropump connected to a stock sample of *E. coli* (10^5 CFU/mL) at 20 ± 5 °C. The SARA loop also included a PBS buffer reservoir on a hot plate maintained at 40 ± 1 °C and a micropump connected to a waste tank. The operating temperature in the hydroponic system was 20 ± 3 °C, which is within the range causing the PNIPAAm nanobrush to extend. The 40 °C PBS buffer was used to actuate nanobrush collapse for measurement.

Upon initiation of the SARA workflow (see supplemental Figure S4), the recirculation pump between the hydroponic tank and the particle trap was turned off. For testing system performance, 10 mL of *E. coli* (ATCC 35218) suspension was injected into a T-junction placed upstream of the particle trap/biosensor apparatus using a microcontroller-operated pump. The concentration of *E. coli* is noted for each addition; stock concentration was 10^5 CFU/mL. An impedance test was conducted at 20 ± 3 °C, and then the program triggered a second pump to rinse the sample with testing buffer at 40 °C (causing the nanobrush to collapse). A second impedance test was triggered by the program, and the data for pre/post actuation is recorded and analyzed. All impedance analysis used a custom handheld potentiostat [19,37]. The code is available by request and was based on two existing codes: ShotBot (https://www.instructables.com/id/ShotBot-Arduino-Powered-Pump-Project/) and Air Piano (https://drive.google.com/file/d/1-E0i1CWxxHpTxtieTizCmZC6Bq6q8Tlc/view).

2.7. Decision Support Application

An artificial reasoning tool (ART) informs the user of the water quality safety according to regulatory standards based on the SARA concept. The decision tree for development of the Thunkable application is shown in supplemental Figure S5. The app design includes screens that contain various functions written in Block code (Scratch). The Thunkable app is publicly available in the searchable open access database (SARA: FC Hydro), or can be provided upon request.

2.8. Statistical Analysis

For all analyses, determinations were made at least in triplicate as independent experiments based on a completely randomized design with equal replications and results were expressed as mean ± standard deviation. Statistical analysis was performed using SPSS PASW Statistics, version 23. Results from the different electrode treatments were tested for significance by analysis of variance (ANOVA) and Tukey's test to separate means at 95% confidence interval ($p < 0.05$).

3. Results

3.1. Electrochemical Characterization

Cyclic voltammograms in 4 mM $Fe(CN)_6^{3-}/Fe(CN)_6^{4-}$ + 1 M KNO_3 (pH = 7.1, T = 25 °C) displayed quasi-reversible redox peaks for all nanomaterials tested, with peak potential separations in the 70 to 80 mV range. Figure 1a shows representative CV curves for bare Pt/Ir electrodes, Pt/Ir electrode after coating with the metal/carbon nanohybrid (nPt-rGO), and the nanohybrid electrode after deposition of nanobrushes (PNIPAAm). Figure 1b was used to calculate the ESA for each electrode. As shown in Figure 1c, the average ESA for nanohybrid electrodes coated with PNIPAAm (0.03 ± 0.004 cm^2) was equivalent to nanohybrid electrodes (0.028 ± 0.002 cm^2), and significantly higher than bare Pt/Ir electrodes (0.018 ± 0.0001 cm^2, $p < 0.05$), similar to the findings reported by Burrs et al. [41].

3.2. Nanobrush Morphology

Electron microscopy and white light interferometry images for the nanocomposite before and after brush deposition are shown in Figure 2. Based on SEM image analysis (Figure 2a), the average particle size for nPt-rGO nanohybrid electrodes was 310 ± 25 nm. SWLI scans over a range of 1 mm × 1 mm show that a homogenous coating was present on the electrode, with heterogeneous nodes that ranged in height from 100 nm to 1 μm. The average profile roughness (R_a) was 166 nm; abrasions on the electrode surface were removed when calculating feature roughness, as these artifacts are a result of repeated polishing with nanodiamond solution during sensor reuse. SEM of the material after nanobrush deposition (Figure 2c) shows terminal nodes of PNIPAAm nanobrushes ranging from 220 to 1300 nm (average node diameter was 910 ± 305 nm). Hills et al. [49] reported the average size of chitosan nanobrush borders contained both terminal nodes (200–300 nm) and longitudinal shafts (100 nm in width and 800 nm in length). However, the shaft structures were not visible in PNIPAAm nanobrushes studied here (Figure 2c, see supplementary Figure S6 for details). This was likely due to the abundance of homogenous terminal nodes on the electrode surface, with shafts forming beneath the nodes (shaft structures are visible in supplementary Figure S6). Electropolymerization of NIPAAM monomers leads to homogeneous nanobrush and formation of uniform shaft and node sizes distribution on the electrode surface. This structure is more of a "brush border" than the chitosan structures by Hills et al. [49], which had a highly heterogenous brush/chain size distribution that likely results from the top-down extraction process (i.e., deacetylation reaction). SWLI (Figure 2d) confirms that PNIPAAm brushes form a homogenous electrode coating at the scale of 800 μm × 800 μm, with an average R_a (201.4 ± 57.2 nm) that was lower than the nPt-rGO hybrid (R_a = 310.2 ± 24.5 nm) material (*t*-test show the reduction in R_a is significant; $p < 0.05$). This reduction in R_a confirms that the PNIPAAm brush structure produces a homogenous electrode coating and provides an excellent platform for conjugation of bioreceptor(s), and many previous papers have shown functionalization of PNIPAAm with a variety of materials [41,50–53]. While we did not test the effect of polymer thickness on performance, the 500–1000 nm thick structures here could be extended to other applications. For example, Zhao et al. [54] developed PNIPAAm-nanocarbon electrodes with thicknesses greater than 5 μm that exhibited high conductivity and electrocatalytic activity but no actuation tests were demonstrated.

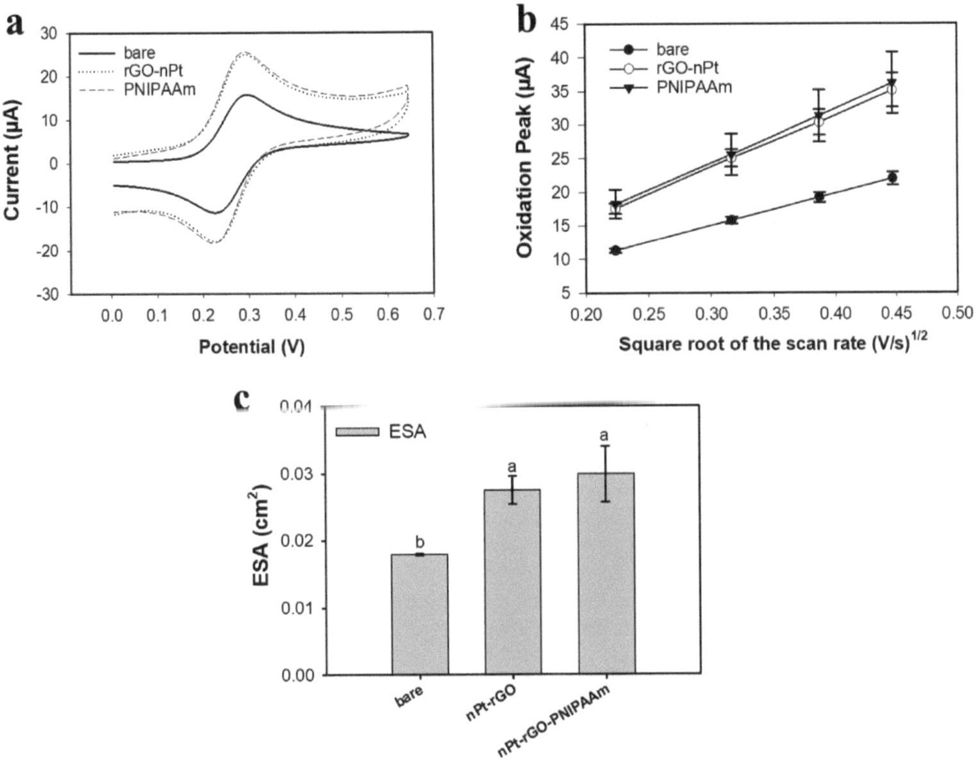

Figure 1. Electrochemical characterization of bare Pt/Ir, nanohybrid (nPt-rGO), and nanohybrid+nanobrush (PNIPAAm) electrodes using 4 mM K_4FeCN_6 as the redox probe. (**a**) Representative CV curves at 100 mV/s scan rate. (**b**) Relationship between I (μA) versus the square root of the scan rate $(V/s)^{1/2}$ for electrode modifications are linear in the range tested. (**c**) Average ESA for various electrode modifications. All data represent the average of three replicates and error bars represent the standard deviation of the arithmetic mean; letters denote significantly different means ($p < 0.05$).

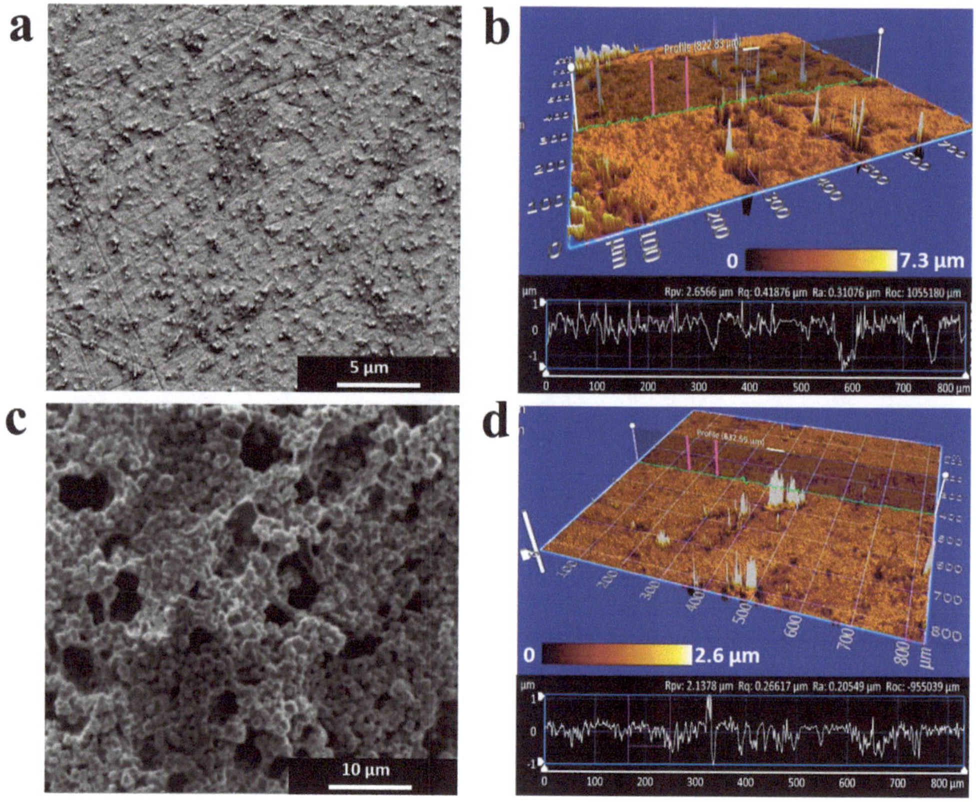

Figure 2. Morphological characterization by scanning electron microscopy (SEM) and scanning white light interferometry (SWLI). (**a**) SEM image for nanohybrid (nPt-rGO) electrode showing homogenous coating with an average feature size of 220 to 1300 nm. (**b**) Average roughness (Ra) for nPt-rGO calculated from SWLI contour profiles (310.2 nm) indicates a heterogenous surface and irregular abrasions. (**c**) SEM images show a relatively homogenous coating for nanobrush-coated nPt-rGO electrodes (PNIPAAm) with terminal nodes 910 ± 305 nm in diameter. (**d**) PNIPAAm-coated electrodes had a more homogenous and smooth coating, with an average Ra of 201.4 nm.

In the next section, we test nanobrush actuation in the presence/absence of *E. coli* and establish optimum conditions for cell capture/impedance testing.

3.3. Nanobrush Actuation and E. coli Capture

The ability to actuate the surface of the sensor with an externally modulated signal (such as local thermodynamic change) is an extremely important property for developing smart sensors that are capable of demonstrating sense-analyze-response-actuate (SARA) behavior. First, ConA and antiGroEL loading onto PNIPAAm nanObrushes was optimized based on ESA calculated from voltammograms in 4 mM K_4FeCN_6 (pH = 7.0, T = 25 °C, see supplementary Figure S7). Protein concentrations between 50–200 nM were tested, and in both experiments the signal saturated for protein concentration of 100 nM (3.2 µM-protein/cm^2), indicating saturation of surface binding sites. Thus, 100 nM was used for all studies in *E. coli* capture.

After conjugation of bioreceptor, the nanobrush detection system was tested using the actuation protocol developed by Hills et al. [49] with some alterations. The original work by Hills et al. [49] targeted *Listeria* spp. using aptamers and antibodies conjugated to chitosan nanobrushes, while in this work *E. coli* is targeted using lectins or antibodies conjugated to

PNIPAAm nanobrushes. In addition to the different bioreceptors used in these studies, the mechanism of actuation for chitosan (a pH-sensitive polymer) is fundamentally different than PNIPAAm (a temperature-sensitive polymer). Thus, *E. coli* capture efficacy was tested above and below the LCST, which is the temperature at which the polymer undergoes a structural change from collapsed (T > LCST) to swollen (T < LCST); the LCST for PNIPAAm is 32–35 °C [50]. The optimum capture strategy was established by analyzing ESA and R_{ct} results for each nanobrush-receptor material.

Figure 3a shows representative CV curves and Figure 3b shows the average charge transfer resistance (R_{ct}) and ESA for nanobrush electrodes at 20 °C and 40 °C using ConA as a receptor (in absence of *E. coli*). The highest redox peaks (Figure 3a) and lowest R_{ct} (Figure 3b) were measured when nanobrushes were collapsed (T = 40 °C > LCST). In the collapsed state, the diffusion layer and the Debye layer are reduced, facilitating higher mass transfer of the electroactive species ($Fe(CN)_6^{3-}/Fe(CN)_6^{4-}$) in this study). This result is similar to the study by Hills et al. [49], where the collapsed state of chitosan led to higher ESA. This clearly shows that PNIPAAm nanobrush actuation is reversible, and changing the polymer from expanded to collapsed is a reliable mechanism for smart sensing.

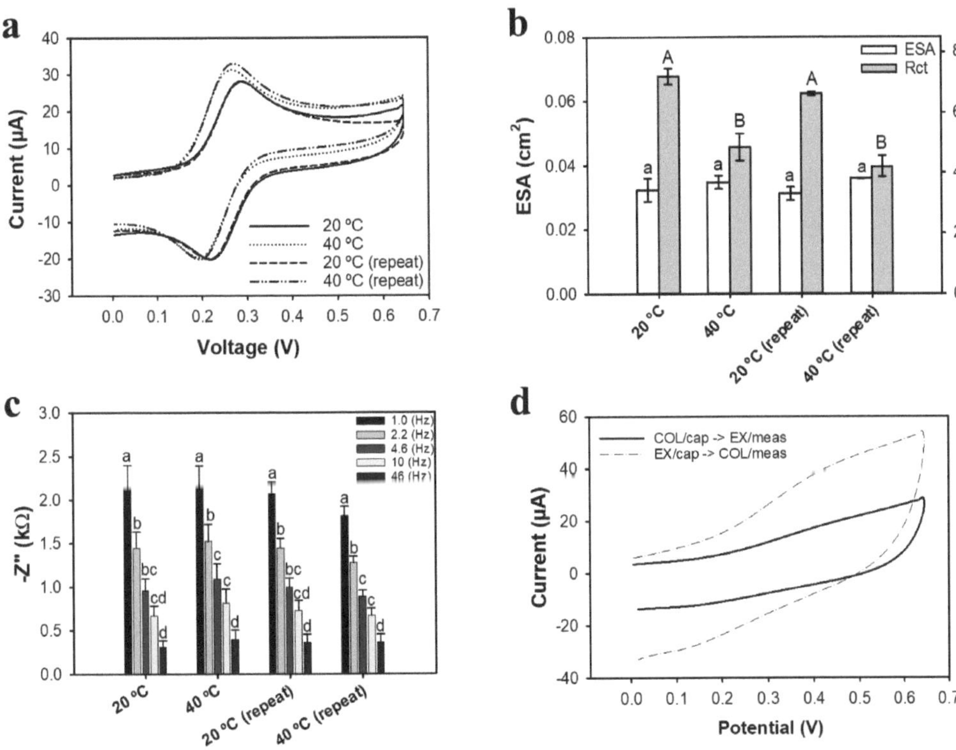

Figure 3. (**a**) Representative CV curves at 100 mV/s. For all tests PNIPAAm brushes were functionalized with ConA (100 nM) and tested at different brush configurations based on temperature response; (**b**) Average charge transfer resistance (R_{ct}) and ESA for different states of PNIPAAm brushes configurations based on temperature response; uppercase and lowercase letters denote significantly different ($p < 0.05$) Rct and ESA means, respectively, for each temperature; (**c**) average impedance at different cutoff frequencies for different temperatures of PNIPAAm actuation; letters denote significantly different means ($p < 0.05$) among the frequencies for each temperature; (**d**) Representative CV (100 mV/s) showing change in peak current for different states of PNIPAAm actuation during *E. coli* capture (10^4 CFU/mL) and measurement. Electrochemical measurements were performed in $Fe(CN)_6^{3-}/Fe(CN)_6^{4-}$ redox probe for (**a–c**), and in PBS for (**d**). Bars represent the average of three replicates.

The maximum impedance values were observed at a cutoff frequency of 1 Hz; analysis considered frequencies from 1 to 50 Hz, which falls within the alpha dispersion domain [64]. No significant difference ($p > 0.05$) was observed between the different temperatures of actuation (Figure 3c). The data in Figure 3c indicate some degree of hysteresis after repeated actuation of bioreceptor-PNIPAAm nanobrushes. After repetitive actuation, peak current changed by 4.0 ± 1.5%, while R_{ct} varied by 13 ± 4%. Repetitive actuation of bioreceptor nanobrush structures has not been reported previously, and warrants further study to determine the best mechanism for actuation.

Two distinguished capture strategies were evaluated using PBS with *E. coli* K12 at a concentration of 1.3 × 10^4 CFU/mL (Figure 3d). First, *E. coli* capture was tested with the nanobrush in the extended state at 20 °C, and impedance measurement was initiated while the brush was in the collapsed state at 40 °C (the opposite condition was also tested). The capture efficiency and signal-to-noise ratio were higher when *E. coli* capture occurred in the extended state (20 °C), followed by brush collapse and subsequent measurement at 40 °C. This finding is consistent with the results reported by Hills et al. [49] and is likely due to higher probability of receptor–*E. coli* interactions when the nanobrush is in an extended state, compared to the collapsed condition where receptor binding site(s) may be inaccessible. Compared to nanobrush actuation without cells (Figure 3a–c), the inverse trend is observed with cell capture followed by sensing (see supplementary Figure S1). During sensing, the nanobrush is collapsed, and the change in signal is likely due to physical crowding of the surface and limited diffusion of redox probes/electrolyte to the surface (a form of Debye shielding).

The concept of material actuation for bacteria capture has also been achieved with magnetically actuated cilia [65–67]. Cilia actuation is used to promote active mixing of bacteria within the unstirred layer, and is not directly involved in the capture mechanism. In the next section, we challenge the biosensor in mixtures of bacteria and establish key performance indicators (KPI) related to sensor engineering.

3.4. Nanobrush Sensing in Buffer

Figure 4 shows PNIPAAm-ConA nanohybrid electrodes calibrated for *E. coli* K12 and in the presence of *Salmonella*. All tests used the actuation protocol previously established (see Figure 3), with capture in the extended state at 20 °C and sensing in the collapsed state at 40 °C. Bode plots are shown over a frequency range of 1 Hz to 100 kHz at varying bacteria concentrations (10^2–10^7 CFU/mL); insets are a zoomed in view of the lower frequency range (1–5 Hz). Based on a frequency analysis using the procedure by Hills et al. [49] and presented in Figure 3c, all calibration curves were developed using data from Bode plots for a cutoff frequency of 1 Hz. The total test time was 17 min, which included 15 min for bacteria capture and 2 min for the EIS measurement.

Figure 4a and supplemental Figure S8a show Bode and Nyquist plots, respectively, for ConA-coated nanobrush sensors targeting *E. coli* K12 in PBS buffer. The average sensitivity toward *E. coli* K12 was 2068.2 ± 346.62 Ω/log(CFU/mL). The LOD (3 sigma, 99.5% confidence interval) for *E. coli* K12 was 5.0 ± 2.9 CFU/mL, and the linear range was 3.0 × 10^2 to 3.0 × 10^5 CFU/mL. The selectivity of the ConA-nanobrush sensor was tested against *Salmonella* Enteritidis (10^2 CFU/mL to 10^7 CFU/mL) in PBS. *Salmonella* Enteritidis was the bacteria chosen for specificity testing due to its similarity to *E. coli* with both being *Gram*-negative and both being a known foodborne pathogen. For a cutoff frequency of 1 Hz, the ConA-nanobrush showed a linear impedance increase for both *E. coli* K12 and *Salmonella* Enteritidis (Figure 4b and supplementary materials Figure S8b, respectively). The average sensitivity toward the mixture of *E. coli* K12 with *Salmonella* Enteritidis was 3800.9 ± 911.2 Ω/log(CFU/mL). The LOD for the mixture was 2.3 ± 0.8 CFU/mL with a linear range from 3.0 × 10^2 to 3.0 × 10^5 CFU/mL.

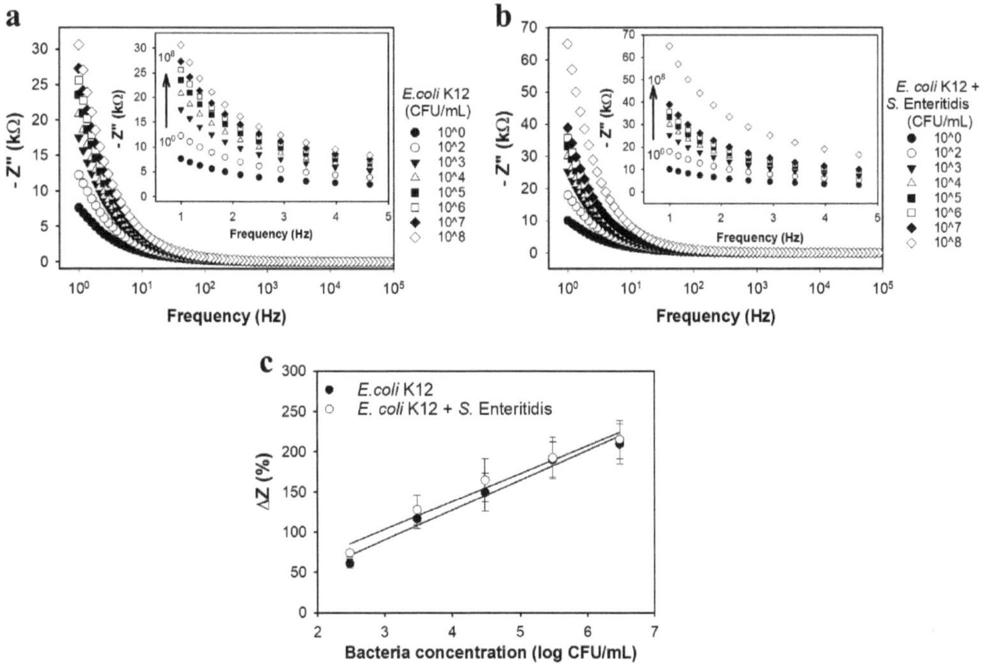

Figure 4. Representative Bode plots for PNIPAAm brush sensors decorated with ConA over the frequency range of 1–100,000 Hz exposed to (**a**) *E. coli* K12 (CFU/mL); and (**b**) equal concentrations of *E. coli* K12 and *Salmonella* Enteritidis (CFU/mL) in PBS, respectively. Insets show exploded view over the frequency range from 1–5 Hz. (**c**) Calibration curves (impedance change at 1 Hz vs. log bacteria concentration) for PNIPAAm-ConA sensors exposed to *E. coli* K12, and *E. coli* K12 and *Salmonella* Enteritidis over their respective linear ranges. All data represents the average of three repetitions. Error bars represent the standard deviation.

Figure 4c presents the linear portion of the calibration curves for ConA biosensors after exposure to *E. coli* or mixtures of *E. coli* with *Salmonella*. ConA biosensors response was similar when exposed to *E. coli* K12 or to mixtures of *E. coli* K12 with *Salmonella* Enteritidis. No significant difference was observed in sensitivity nor LOD for sensors exposed to *E. coli* K12 or mixtures of *E. coli* K12 with *Salmonella* Enteritidis. *E. coli* and *Salmonella* are both *Gram*-negative enterobacteria, but differ in sugar combining sites on the cell surface and LPS (lipopolyssacharides) structures, each affected by growth state [68–70]. *E. coli* K12 LPS is terminated by O-antigen, which contains glucose that has a high affinity for ConA [71]. The specific binding site for ConA on *E. coli* is thought to be G_{M1} ganglioside [72], while the specific binding site on *Salmonella* varies since *Salmonella* express a range of LPS with dynamic numbers of terminal O antigen [73].

The presence of *Salmonella* Enteritidis in the testing solution did not show significant interference ($p > 0.05$) on the slope of the calibration curve (Figure 4c), indicating no cross-reaction between the PNIPAAm-ConA sensor and the *Salmonella* Enteritidis. *E. coli* and *Salmonella* both have lipopolysaccharides on their cell membranes that are terminated with carbohydrates that may bind ConA. The established theory is that the carbohydrate-recognition domain of ConA forms an association with the terminal sugar of microbial LPS structures through an interaction with both the 3- and 4-OH groups of the sugar. As reviewed by Vanegas et al. [2], the molecular interaction is more complex than a simple protein–ligand interaction. The terminal sugar does not always determine ConA binding, and the lectin is somewhat promiscuous and can recognize branch sugars deeper within the LPS. This is an important feature when considering the actuation of ConA-nanobrushes for improving capture of *Gram*-negative bacteria, as the three-dimensional structure may be crucial to binding, also reaffirmed by the Spike protein of SARS-CoV-2 [74–76].

In the next section, we challenge the biosensor against the pathogen *E. coli* O157:H7 in food samples for both lectin-terminated nanobrushes and antibody-terminated brushes.

3.5. Nanobrush Sensing in Food Samples

ConA-nanobrush and Ab-nanobrush (Anti-GroEL) biosensors were tested with *E. coli* O157:H7 at concentrations ranging from 10^2 to 10^8 CFU/mL in sterile vegetable broth using the actuation protocol described above. Figure 5a,b present Bode plots over a frequency range of 1 Hz to 100 kHz for ConA-nanobrush and Ab-nanobrush biosensors, respectively (see supplementary Figure S9 for Nyquist plots).

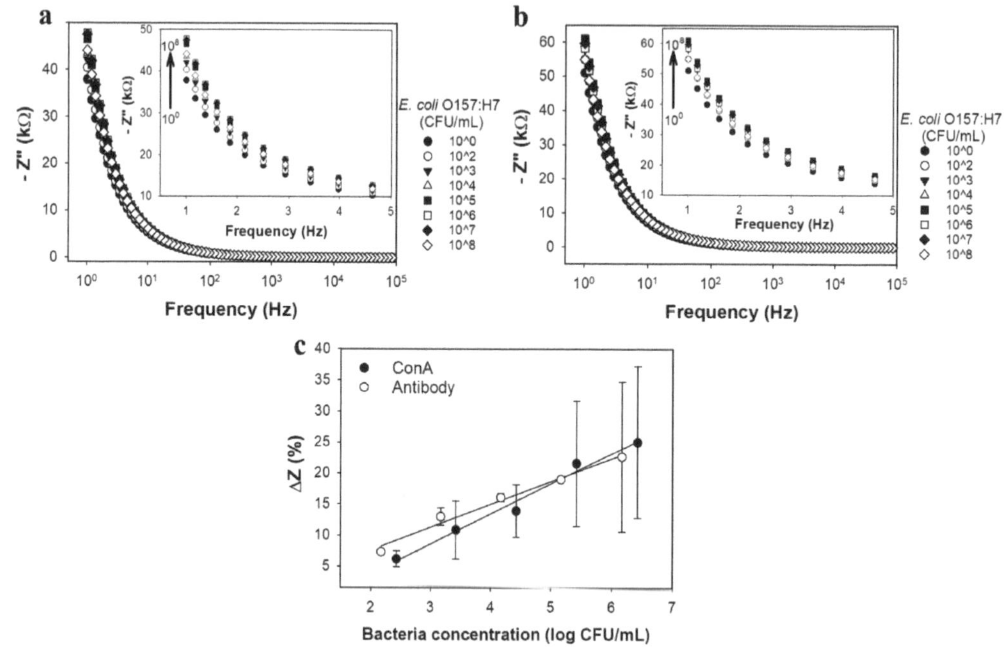

Figure 5. Representative Bode plots over the frequency range of 1–100,000 Hz for PNIPAAm brush sensor decorated with (**a**) ConA; and (**b**) Anti-GroEL antibody exposed to various concentrations of *E. coli* O157:H7 (CFU/mL) in vegetable broth. Insets show the exploded view over the frequency range of 1–5 Hz. (**c**) Calibration curves at 1 Hz for PNIPAAm brush sensor decorated with ConA and Anti-GroEL antibody exposed to *E. coli* O157:H7 in vegetable broth over their respective linear ranges. All data represents the average of three repetitions. Error bars represent standard deviation.

Figure 5c shows the calibration curves for each receptor in vegetable broth based on net impedance. Range and lower LOD obtained from the linear portions of the calibration curves for PNIPAAm-ConA and PNIPAAm-antibody sensors are shown in Table 1. The sensitivity (slope of the calibration curves in Figure 5c) of both nPt-rGO-PNIPAAm-ConA (1915.0 $\pm$ 1070.1 Ω/log(CFU/mL)) and nPt-rGO-PNIPAAm-Anti-GroEL antibody (2004 $\pm$ 253.6 Ω/log(CFU/mL)) sensors exposed to *E. coli* O157:H7 is similar ($p > 0.05$), indicating the potential of both in the detection of *E. coli* O157:H7 in a complex system. The LOD for the PNIPAAm-antibody (249.3 $\pm$ 51.3 CFU/mL) was significantly lower ($p < 0.05$) than the PNIPAAm-ConA (1560 $\pm$ 202.3 CFU/mL); however, the coefficient of variation (CV) for PNIPAAm-ConA (0.13) was lower than the PNIPAAm-antibody (0.21), indicating less variability in the results obtained for the former sensor. However, both sensitivity values were on the same order of magnitude and the linear ranges were similar, which supports the hypothesis that both are capable of *E. coli* O157:H7 detection.

Table 1 contains a compilation of current biosensors for the detection of *E. coli* in various food samples, or buffers. The detection time for our actuating nanobrush sensor was shorter than all times found in the literature, except for the sensor used by Radke and Alocilja [77], who reported a 10 min detection time. The LOD of the nanobrush sensors for *E. coli* K12 in PBS (5.0 ± 2.9 CFU/mL) was lower than all biosensors shown in the table. Although the LODs of the nanobrush sensors for *E. coli* O157:H7 were significantly higher (1560 ± 202.3 CFU/mL for ConA and 249.3 ± 51.3 CFU/mL for antibody), these LODs are within the same order of magnitude of previous studies. Additionally, their sensitivity in vegetable broth (1915.0 ± 1070.1 Ω/log(CFU/mL)) was comparable to the test in PBS for *E. coli* K12 (2068.2 ± 346.62 Ω/log(CFU/mL)). The complex composition of vegetable broth can promote non-specific interactions and influence the electrochemical response Hills et al. [49]. Complex media, including whole milk, lettuce wash water, and ground beef, are usually used to measure the performance of sensors in real food samples [18,35,41,78]. In this study, vegetable broth was chosen, as it is composed of food ingredients with the potential for contamination by foodborne pathogens, namely *E. coli* O157:H7. The performance of the PNIPAAm nanobrush biosensors indicates that the devices have potential to be used in similar complex solutions, such as fresh produce wash water and other aqueous media. A significant advantage of the biosensors in this work is the lack of bacteria purification or concentration steps—see, for example, Chowdhury et al. [79]—that require label addition and incubation. This is a critically important feature for SARA biosensors, as real time analysis using remote system actuation is a defining feature of the tool. In the next section, we show how this nanobrush biosensor can be used to create a SARA system for rapid analysis of lettuce irrigation water.

3.6. SARA: Sense-Analyze-Respond-Actuate

Figure 6 shows the system used for a proof of concept SARA demonstration, highlighting actuation at the macroscale (pump state switching via microcontrollers), and actuation at the nanoscale (polymer swelling/contraction for bacteria capture). Detailed schematics may be found in the Supplementary Section.

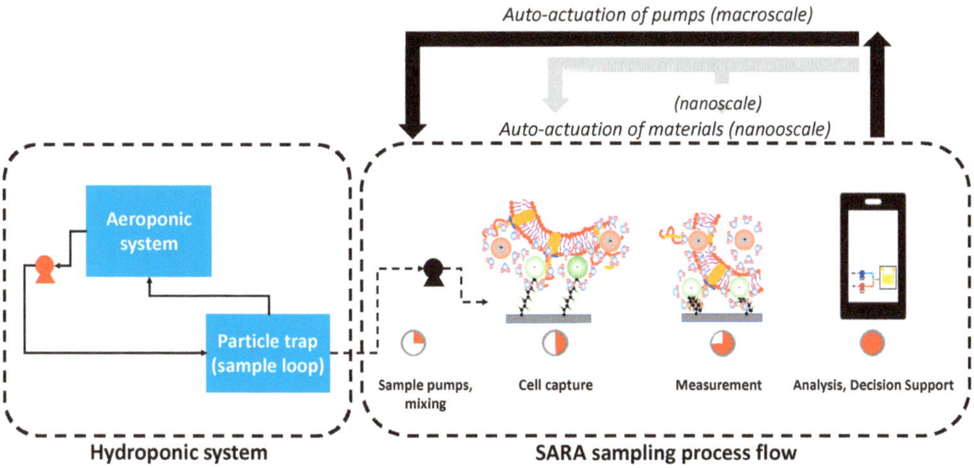

Figure 6. Simplified schematic of SARA system, highlighting actuation at the macroscale (microcontrollers used to activate/deactivate pumps) and the nanoscale (polymer swelling/collapse during bacteria capture).

Table 1. Biosensor performance comparison to other devices in the literature used in the detection of *E. coli* O157:H7.

Sensing Platform/Biorecognition Agent	Media	Bacteria	Detection Time (min)	Range (CFU/mL)	LOD (CFU/mL)	Reference
PNIPAAm-ConA	PBS	*E. coli* K12	17	3×10^2 to 3×10^5	5.0 ± 2.9	This work
PNIPAAm-ConA	PBS	*E. coli* K12 and *Salmonella* Enteritidis	17	3×10^2 to 3×10^5	2.3 ± 0.8	This work
PNIPAAm-ConA	Vegetable broth	*E. coli* O157:H7	17	2.6×10^2 to 2.6×10^6	1560.0 ± 202.3	This work
PNIPAAm-Antibody	Vegetable broth	*E. coli* O157:H7	17	1.5×10^2 to 1.5×10^6	249.3 ± 51.3	This work
Polyanilyne-Antibody	Phosphate Citrate Buffer	*E. coli* O157:H7	NR	10^2 to 10^7	10^2	Chowdhury et al. [80]
11-MUA SAM-ConA *	Carrier buffer	*E. coli*	<20	12 to 1.2×10^6	12	Jantra et al. [81]
Hyaluronic acid-Antibody	PBS	*E. coli* O157:H7	NR	10 to 10^5	10	Joung et al. [78]
Hyaluronic acid-Antibody	Whole milk	*E. coli* O157:H7	NR	NR	83	Joung et al. [78]
CHIT-MWNTs-SiO2@THI **	PBS	*E. coli* O157:H7	<45	4.1×10^2 to 12×10^5	250	Li et al. [82]
Bayhydrol 110-Mannose-ConA ***	Water	*E. coli* O157:H7	180	6 to 60×10^9	60	Lu et al. [83]
Gold electrode array-Antibody	Lettuce Wash Water	*E. coli* O157:H7	10	10^4 to 10^7	10^4	Radke & Alocilja [77]

Values provided are means of three replicates ± standard deviations. Means that are not followed by a common superscript letter are significantly different ($p < 0.05$). NR denotes values not reported. * *E. coli* strain not specified in literature. ** Chitosan-multiwalled carbon nanotubes-SiO_2/thionine. *** Mannose immobilized onto the polymer and ConA added on sample solution for the biorecognition process.

Figure 7 shows the average data from repetitive tests of the SARA system (raw impedance data are available upon request). Figure 7a shows the average data for all tests, with colored bar charts indicating equivalent bacteria concentration. The three successive values represent impedance data obtained at 20 °C (extended nanobrush), 40 °C (collapsed nanobrush), and return to 20 °C (extended brush + bacteria). In the absence of bacteria, the R_{ct} decreased when nanobrushes were in the collapsed state, while in the presence of bacteria, this trend was reversed. These system-scale measurements with a handheld potentiostat confirm the bench scale measurement in Figure 3. For all bacteria concentrations, the percent change in R_{ct} during actuation (18.4 ± 6.3%) was significantly higher than the tests without bacteria (13 ± 4%). Interestingly, the percent change in R_{ct} decreased with increasing *E. coli* concentration, which was counter-intuitive when comparing data to baseline measurements, see supplementary Figure S11. Figure 7b shows the average sensor calibration using an extend–capture, collapse–measure scheme.

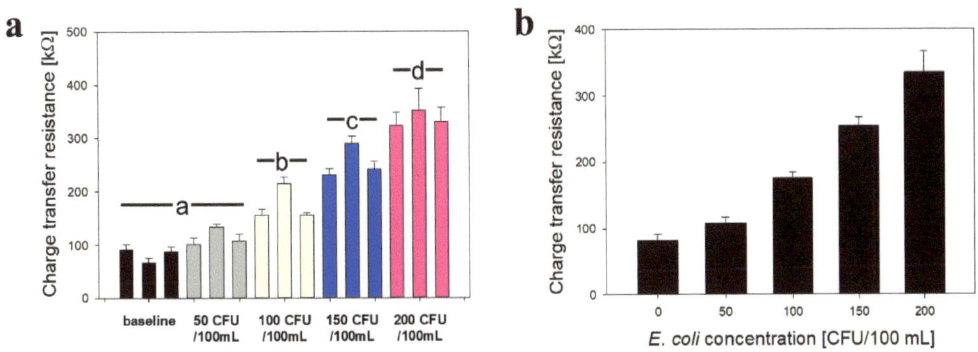

Figure 7. Replicate impedance data for actuation of *E. coli* biosensors in hydroponic lettuce system. (**a**) Average charge transfer resistance for all tests, with colored bar charts indicating equivalent bacteria concentration. The three successive values represent impedance data obtained at 20 °C (extended nanobrush), 40 °C (collapsed nanobrush), and return to 20 °C (extended brush + bacteria). (**b**) Average sensor calibration using using an extend–capture, collapse–measure scheme. For simulating contamination, 10 mL of an *E. coli* (ATCC 35218) stock suspension was injected into a T-junction placed upstream of the particle trap/biosensor apparatus. All data represents the average of three repetitions. Error bars represent standard deviation. Different letters denote significantly different means ($p < 0.05$)

The average sensitivity toward *E. coli* was 366 ± 81 Ω/log(CFU/mL), which is lower than the laboratory assay by nearly two-fold. However, SARA is an in situ system and does not require sample extraction and transport to an analytical laboratory. Furthermore, the acquisition system is based on an open source smartphone tool [37] and enables access to users who do not have specialty equipment. The LOD (3 sigma; 99.5% confidence interval) for *E. coli* was 57.6 ± 13.4 CFU/mL, and the linear range was 50 to 200 CFU/mL. In the next section, we show development and use of a smartphone application for providing on site decision support, closing the analysis loop for SARA.

3.7. Decision Support Application

A decision support app was developed for iOS using Thunkable (block coding). The aim was to provide meaningful information from the real-time analysis of sensor data that meets statistical thresholds, and subsequently uses guidance from PSR to determine whether the sample meets regulatory compliance. Figure 8 shows select screenshots from the app, which is a tool within the SNAPS (sensor analytic point solutions) portfolio, which uses ART (artificial reasoning tools) [19]. After two-factor authentication and authorization via DUO [84], the welcome screen contains an option for enabling GPS (location accuracy), which is an important step if location-specific decision support is desired. The first-generation application uses the federal guidance for water quality from the PSR, but the application can be expanded to produce locally relevant guidance, if required. The REST architecture [85] facilitates expansion for linking to other regulatory databases through open RESTful APIs (in synchronous or asynchronous mode). When available, batch calibration may be enabled using QR codes embedded with calibration data (the app also contains an option for manual calibration). Error due to batch calibration can be as high as 5% due to batch-to-batch variation. Screen number 5 provides the option to connect to a sensor reader via Bluetooth (or WiFi, WiMax), or enter data manually (may be error prone). A decision support option screen can provide location-specific decision support (if GPS is enabled and databases are available), or generic decision support. The final screen provides basic information regarding the functionality of the decision support and disclaimers as relevant. The protocol for development of the app can be found in the supplemental section. Use of Block Coding, originally developed as Scratch coding [75], and the open-source platform Thunkable ensure that others may replicate the tool.

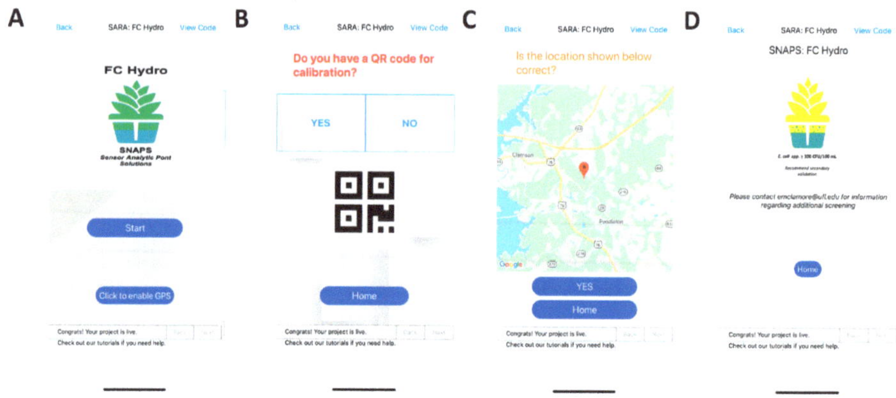

Figure 8. Select screenshots of SNAPS FC Hydro decision support app. (**A**) Welcome screen with option for location accuracy. (**B**) QR code scanner option for batch calibration input; (**C**) Location accuracy for decision support; (**D**) Example of result indicating sample requires further processing and may be contaminated with *E. coli* (>100 CFU *E. coli*/100 mL).

3.8. Decision Support Application—Cybersecurity

Any discussion of actuation or automation, even if it is partial, must recognize the possibility of cyberthreats. Cybersecurity, by design, is required for implementing the SARA paradigm. SNAPS LM Hydro decision support system (DSS) depends on connectivity using open tools such as IFTTT [86] (if this, then, that) or may integrate with time sensitive networking [87], which creates opportunities for cyber-disruption. Digital by design embraces the concept of IoT [88], and elements of cybersecurity [89] are essential to its core design rather than an "after-thought" in the execution layer. The design of cybersecurity is beyond the scope of this paper. We touch upon a few common cyberthreat scenarios and suggest commercial tools to mitigate risks. Cybercrimes may involve: (i) time alteration, (ii) GPS spoofing, (iii) intruders, and iv) embedded systems.

(i) Time alteration: Elementary cyberthreat which may precipitate discord simply by altering the time when an action is performed (actuated).

Time jamming targets the network time protocol (NTP) by amplifying [90] requests to a NTP server (distributed denial of service [91] or DDoS attack) which makes time "unavailable" to the command execution layer. Internet Engineering Task Force's (IETF) manufacturer usage description (MUD) architecture is "supposed" to enable (IoT [92]) devices to send and receive only the traffic they require to perform as intended (network will prohibit other communication with the devices).

Time spoofing works by providing the wrong time, for example, changing the time in reverse such that an event can be accomplished in that interval or execute an event using an expired (timestamp) authorization key or code which can be reused if the system accepts it as "valid" because the "spoofed" time is still within the valid time window. Problems with time are compounded in software-driven commands to execute physical processes (cyberphysical systems, programmable logic controllers) not only due to engineering constraints [93] (bandwidth, latency, jitter) but because the semantics of time and the physical notion of "timeliness" is still not a part of software architecture [94]. Perhaps this dysfunctional gap in time is due to intrinsic human inability to grasp the epistemology of temporal semantics. To deploy SARA, we have to address cyberthreats in the context of the SNAPS LM Hydro DSS because delayed actuation could change system performance. Vendors provide a variety of tools which may be integrated with SNAPS DSS cloud as a risk mitigation strategy (in consultation with customers/users).

(ii) GPS spoofing: Incorrect location (GPS spoofing [95]) of sensor data could destabilize food safety and security. For example, SARA-induced auto-actuation may release contaminated wastewater to flood an already moist (water saturated) field growing produce or redirect a false alarm about pathogens (causing humans in the loop to discard food or produce). With increasing adoption of SARA/SNAPS, the risks [96] for open-source CPS [97] projects due to Global Navigation Satellite System (GNSS) spoofing may result in fake position, navigation, and time (PNT) information. The latter may cause economic loss and social unrest.

The relative ease [98] of GPS spoofing with software-defined radio (SDR) is due to the fact that GPS transmits an open signal without any encryption, which can be recorded, manipulated and transmitted. GPS location spoofing often converges with time spoofing (PNT [99]) because the GPS network of satellites broadcast time-stamped messages continually. The hacker may alter how long (time) it takes for the signals from each satellite to reach the receiver (multiplying this time by the speed of light gives the distance between the receiver and each satellite). Interference detection and mitigation (IDM) tools may be integrated with SARA/SNAPS to protect [100] against overt and covert [101] GPS spoofing, according to project requirements.

(iii) Intruders: Security of applications against intruders and intruder detection [102] is an immense and dynamic problem in several domains including IoT scenarios. In the context of SARA/SNAPS, it is important to strike an operational balance between the infinite breadth of this problem versus a feasible inclusion of a verification system which can authenticate and authorize users accessing various parts of the SNAPS network/application. We propose using multi-factor authentication [103] (MFA) as a tool to verify the identity of users using a code, token or certificate usually linked via a mobile app to doubly authenticate the user during the sign-in process. In addition, MFA may also use a re-validation step using a remotely stored certificate [104] in a home institution [105] or on a trusted third-party server, or use biometrics, such as a fingerprint or retina scan.

(iv) Embedded systems: Emergence of the networked physical world made it clear that any form of identification [106] (objects, processes, humans) must address security [107] implications. Extending the IoT metaphor to SARA/SNAPS implies that the nature of messaging between components (sensors, devices, gateways) is vulnerable to intrusion. Enhancing security [108] within the SARA/SNAPS ecosystem may avoid commercial excesses but explore whether tools like MAM [109] or masked (encrypted) authenticated (device verification) messaging (data) may be feasible for deployment. Thus, in combination, MFA and MAM provide authentication, at least partially, for users, data and devices, in our attempt to address a basic layer of cybersecurity.

4. Conclusions

Current standard methods of foodborne pathogen detection including culture and colony counting, ELISA, and PCR require training and are time consuming. As foodborne pathogens are a persistent concern in the food industry, there is a demand for a rapid, reliable, and cost-effective detection method. The biosensors developed here using 100 nM of Concanavalin A (ConA) lectin or antibody in a PNIPAAm nanobrush sensing platform achieved a rapid detection (17 min total) of *Escherichia coli* in a buffer solution (PBS), as well as in a real-world scenario simulated by a complex vegetable broth. The use of thermo responsive polymer brush interfaces in combination with hybrid nPt-rGO nanostructures were shown to enhance the capture of target *E. coli* bacteria and transduction of electrochemical outputs as the acquisition signal. The optimum conditions were to capture bacteria at 20 °C when PNIPAAm nanobrushes were expanded, and initiate the test sequence at 40 °C when the brushes were collapsed. Expanded brushes permitted biorecognition agents to more easily attach to bacteria, while the collapsed state assisted in electrochemical response. The nPt-rGO-PNIPAAm-ConA sensor presented significantly low LOD (LOD of 5.0 ± 2.9 CFU/mL), high sensitivity (2068.2 ± 346.6 Ω/log(CFU/mL) and selectivity (LOD of 2.3 ± 0.8 CFU/mL) to *E. coli* on the performance tests with *E. coli* K12 alone and also with *E. coli* K12 and *Salmonella* Enteritidis in the PBS, respectively. Selectivity to *E. coli* in the presence of *Salmonella* was important in evaluating the success of the sensor and comparison to preexisting sensors. On the tests in vegetable broth inoculated with *E. coli* O157:H7 to mimic a real-world, complex sample in the presence of a foodborne pathogen, the PNIPAAm-ConA sensor showed sensitivity (1915.0 ± 1070.1 Ω/log(CFU/mL)) similar ($p > 0.05$) to the PNIPAAm-antibody sensor (2004 ± 253.6 Ω/log(CFU/mL)). Hence, the application of the nPt-rGO-PNIPAAm as a platform for both ConA and antibodies provided superior performance results when detecting *E. coli* in pristine media such as PBS as well as in a complex system such as vegetable broth compared to the literature. The advantages of ConA over antibodies in terms of production, cost, and shelf-life combined with comparable results to other biosensors make nPt-rGO-PNIPAAm-ConA a potential alternative to current detection methods for testing food samples.

Extending this biosensor application and potential adoption, SARA represents a rapid label-free bacteria detection system for use in agricultural waters where nanomaterial actuation is triggered by a smartphone. Actuation of the nanobrush takes place without requiring the user to perform any additional tasks. The decision support tool (ART) is embedded in the smartphone used for signal acquisition. Our use of polymer nanobrushes in biosensing significantly advances the field by developing a new cyber-physical tool which is amenable to direct control over the brush border structure, rather than relying on specialty equipment within a laboratory. SARA extends the spatial resolution in processing facilities to contain the spread of contamination. Such detection methods raise even more interest when the economic losses due to recalls of contaminated food or the extra time to release the product are considered.

Supplementary Materials: The following are available online at https://www.mdpi.com/2076-0825/10/1/2/s1: Figure S1. Time course of SARA testing system. Sample pumps provide mixing, sampling, and temperature control. Figure S2. Cyclic voltammetry of the SPC functionalized electrodes before and after freeze-drying process. Figure S3. Simplified process flow diagram showing semi-closed loop batch aeroponic system and SARA biosensor loop. Figure S4. SARA workflow for programming the microcontroller. Figure S5. Decision tree for design of decision support application. Figure S6. SEM images of nPt-rGO electrodes with PNIPAAM nanobrushes (a) top view and (b,c) cross-sectional view at 10 kV and 25,000, 10,000, and 29,000 times magnification, respectively. Figure S7. Comparison of ESA in cm^2 and representative CV curves at 100 mV/s for nPt-rGO coated electrodes at different (a,b) antibody concentrations; and (c,d) ConA concentrations Figure S8. Representative Nyquist plots for PNIPAAm brush sensors decorated with ConA over the frequency range of 1–100,000 Hz exposed to (a) *E. coli* K12 (CFU/mL), and (b) equal concentrations of *E. coli* K12 and *Salmonella* Enteritidis (CFU/mL) in PBS. Figure S9. Representative Nyquist plots over the frequency range of 1–100,00 Hz for PNIPAAm brush sensors decorated with (a) ConA; and (b) Anti-GroEL antibody exposed to various concentrations of *E. coli* O157:H7 (CFU/mL) in vegetable broth. Figure S10. Representative Bode plots over the frequency range of 1–100,000 Hz for (a) nPt-rGO-ConA and (b) nPt-rGO-Anti-GroEL antibody modified electrodes exposed to various concentrations of *E. coli* K12 (CFU/mL) in PBS. Figure S11. Percent change in R_{ct} during SARA system testing. The percent change decreased significantly with increasing *E. coli* concentration.

Author Contributions: Conceptualization, S.P.A.D., E.S.M. and C.L.G.; methodology, C.A.G., D.A.O., C.C.P., D.X., Y.T., E.S.M. and C.L.G.; software, D.X., S.P.A.D., E.S.M.; validation, C.A.G., D.A.O., D.X., Y.T. and E.S.M.; formal analysis, C.A.G., D.A.O., C.C.P, E.S.M. and C.L.G.; resources, E.S.M. and C.L.G.; data curation, E.S.M. and C.L.G.; writing—original draft preparation, C.A.G., E.S.M. and C.L.G.; writing—review and editing, D.A.O., C.C.P., S.P.A.D., E.S.M. and C.L.G.; visualization, D.A.O., C.C.P., S.P.A.D., E.S.M. and C.L.G.; supervision, E.S.M. and C.L.G.; funding acquisition, E.S.M. and C.L.G. All authors have read and agreed to the published version of the manuscript.

Funding: This project was supported by Agriculture and Food Research Initiative (AFRI) Competitive Grant no. 2018-67016-27578 awarded as a Center of Excellence from the USDA National Institute of Food and Agriculture, the National Science Foundation Nanobiosensors program (Grants No. CBET-1511953/1512659), and USDA-AFRI project number 2017-08795. The authors also acknowledge support from the UF IFAS Equipment Grant (ESM). A pre-print of this paper is in the MIT Library and may be downloaded using this URL: https://dspace.mit.edu/handle/1721.1/123983.

Institutional Review Board Statement: Not applicable.

Informed Consent Statement: Not applicable.

Data Availability Statement: The data presented in this study are available on request from the corresponding author.

Conflicts of Interest: The authors declare no conflict of interest.

References

1. Valderrama, W.B.; Dudley, E.G.; Doores, S.; Cutter, C.N. Commercially Available Rapid Methods for Detection of Selected Food-borne Pathogens. *Crit. Rev. Food Sci. Nutr.* **2016**, *56*, 1519–1531. [CrossRef]
2. Vanegas, D.C.; Gomes, C.L.; Cavallaro, N.D.; Giraldo-Escobar, D.; McLamore, E.S. Emerging Biorecognition and Transduction Schemes for Rapid Detection of Pathogenic Bacteria in Food. *Compr. Rev. Food Sci. Food Saf.* **2017**, *16*, 1188–1205. [CrossRef]
3. Wang, Y.; Salazar, J.K. Culture-Independent Rapid Detection Methods for Bacterial Pathogens and Toxins in Food Matrices. *Compr. Rev. Food Sci. Food Saf.* **2016**, *15*, 183–205. [CrossRef]
4. Umesha, S.; Manukumar, H.M. Advanced molecular diagnostic techniques for detection of food-borne pathogens: Current applications and future challenges. *Crit. Rev. Food Sci. Nutr.* **2018**, *58*, 84–104. [CrossRef]
5. Centers for Disease Control and Prevention (CDC). National Outbreak Reporting System (NORS). 2020. Available online: https://wwwn.cdc.gov/norsdashboard/ (accessed on 14 October 2020).
6. Centers for Disease Control and Prevention (CDC). Estimates of Foodborne Illness in the United States. 2018. Available online: https://www.cdc.gov/foodborneburden/2011-foodborne-estimates.html (accessed on 27 July 2020).
7. Rai, P.K.; Tripathi, B.D. Microbial contamination in vegetables due to irrigation with partially treated municipal wastewater in a tropical city. *Int. J. Environ. Health Res.* **2007**, *17*, 389–395. [CrossRef]
8. Callejón, R.M.; Rodríguez-Naranjo, M.I.; Ubeda, C.; Hornedo-Ortega, R.; Garcia-Parrilla, M.C.; Troncoso, A.M. Reported Foodborne Outbreaks Due to Fresh Produce in the United States and European Union: Trends and Causes. *Foodborne Pathog. Dis.* **2015**, *12*, 32–38. [CrossRef]
9. Odonkor, S.T.; Ampofo, J.K. *Escherichia coli* as an indicator of bacteriological quality of water: An overview. *Microbiol. Res.* **2013**, *4*, e2. [CrossRef]
10. Feng, P.; Weagant, S. Chapter 4A: Diarrheagenic *Escherichia coli*. U.S. Food & Drug Administration (FDA). Bacteriological Analytical Manual. 2011. Available online: https://www.fda.gov/food/laboratory-methods-food/bam-chapter-4a-diarrheagenic-escherichia-coli (accessed on 24 June 2020).
11. Walker, C.L.F.; Applegate, J.A.; Black, R.E. Haemolytic-uraemic syndrome as a sequela of diarrhoeal disease. *J. Health Popul. Nutr.* **2012**, *30*, 257–261. [CrossRef] [PubMed]
12. Rasko, D.A.; Rosovitz, M.J.; Myers, G.S.A.; Mongodin, E.F.; Fricke, W.F.; Gajer, P.; Crabtree, J.; Sebaihia, M.; Thomson, N.R.; Chaudhuri, R.; et al. The Pangenome Structure of *Escherichia coli*: Comparative Genomic Analysis of *E. coli* Commensal and Pathogenic Isolates. *J. Bacteriol.* **2008**, *190*, 6881–6893. [CrossRef] [PubMed]
13. Centers for Disease Control and Prevention (CDC). Multistate Outbreak of *E. coli* O157:H7 Infections Linked to Romaine Lettuce. 2018. Available online: https://www.cdc.gov/ecoli/2018/o157h7-04-18/index.html (accessed on 24 July 2020).
14. Allende, A.; Datta, A.R.; Smith, W.A.; Adonis, R.; MacKay, A.; Adell, A.D. Implications of new legislation (US FSMA) and guidelines (EC) on the establishment of management systems for agricultural water. *Food Microbiol.* **2018**, *75*, 119–125. [CrossRef]
15. U.S. Food and Drug Administration (FDA). *FSMA Final Rule on Produce Safety*; FDA: Silver Spring, MD, USA, 2020.
16. U.S. Food and Drug Administration (FDA). FSMA Rules & Guidance for Industry. 2020. Available online: https://www.fda.gov/food/food-safety-modernization-act-fsma/fsma-rules-guidance-industry (accessed on 24 July 2020).

17. Rezaei, A. Food safety: The farmer first health paradigm. *One Health* **2018**, *5*, 69–73. [CrossRef] [PubMed]
18. Ahearn, M.C.; Armbruster, W.; Young, R. Big Data's Portential to Improve Food Supply Chain Environmental Sustainability and Food Safety. *Int. Food Agribus. Manag. Rev.* **2016**, *19*, 155–171.
19. McLamore, E.S.; Palit Austin Datta, S.; Morgan, V.; Cavallaro, N.; Kiker, G.; Jenkins, D.M.; Rong, Y.; Gomes, C.; Claussen, J.; Vanegas, D.; et al. SNAPS: Sensor Analytics Point Solutions for Detection and Decision Support Systems. *Sensors* **2019**, *19*, 4935. [CrossRef] [PubMed]
20. Armbruster, W.; MacDonell, M.M. Informatics to Support International Food Safety. In Proceedings of the 28th EnviroInfo 2014 Conference, Oldenburg, Germany, 10–12 September 2014.
21. Marvin, H.J.P.; Janssen, E.M.; Bouzembrak, Y.; Hendriksen, P.J.M.; Staats, M. Big data in food safety: An overview. *Crit. Rev. Food Sci. Nutr.* **2017**, *57*, 2286–2295. [CrossRef] [PubMed]
22. Food and Drug Administration (FDA). Standards for the Growing, Harvesting, Packing, and Holding of Produce for Human Consumption: What You Need to Know About the FDA Regulation: Guidance for Industry Small Entity Compliance Guide. 2017. Available online: https://www.fda.gov/media/107298/download (accessed on 24 July 2020).
23. Rock, C.M.; Brassill, N.; Dery, J.L.; Carr, D.; McLain, J.E.; Bright, K.R.; Gerba, C.P. Review of water quality criteria for water reuse and risk-based implications for irrigated produce under the FDA Food Safety Modernization Act, produce safety rule. *Environ. Res.* **2019**, *172*, 616–629. [CrossRef]
24. Markland, S.M.; Ingram, D.; Kniel, K.E.; Sharma, M. Water for Agriculture: The Convergence of Sustainability and Safety. *Microbiol. Spectr.* **2017**, *5*. [CrossRef]
25. United States Department of Agriculture Food Safety and Inspection Services (USDA FSIS). FSIS Compliance Guideline for Controlling Meat and Poultry Products Pending FSIS Test Results. 2013. Available online: https://www.fsis.usda.gov/wps/wcm/connect/6b7b5a65-9ad3-4f89-927d-078e564aca24/Compliance_Guide_Test_Hold_020113.pdf?MOD=AJPERES (accessed on 24 July 2020).
26. Pozo, V.F.; Schroeder, T.C. Evaluating the costs of meat and poultry recalls to food firms using stock returns. *Food Policy* **2016**, *59*, 66–77. [CrossRef]
27. Castell-Perez, E.; Gomes, C.; Tahtouh, J.; Moreira, R.; McLamore, E.S.; Knowles, H.S. Food Processing and Waste Within the Nexus Framework. *Curr. Sustain. Renew. Energy Rep.* **2017**, *4*, 99–108. [CrossRef]
28. Keener, L.; Nicholson-Keener, S.M.; Koutchma, T. Harmonization of legislation and regulations to achieve food safety: US and Canada perspective. *J. Sci. Food Agric.* **2014**, *94*, 1947–1953. [CrossRef]
29. Morgan, V.; Casso-Hartmann, L.; Bahamon-Pinzon, D.; McCourt, K.; Hjort, R.G.; Bahramzadeh, S.; Velez-Torres, I.; McLamore, E.; Gomes, C.; Alocilja, E.C.; et al. Sensor-as-a-Service: Convergence of Sensor Analytic Point Solutions (SNAPS) and Pay-A-Penny-Per-Use (PAPPU) Paradigm as a Catalyst for Democratization of Healthcare in Underserved Communities. *Diagnostics* **2020**, *10*, 22. [CrossRef]
30. Vanegas, D.; Gomes, C.; McLamore, E. Biosensors for Indirect Monitoring of Foodborne Bacteria. *Biosens. J.* **2016**, *5*, 137–139. [CrossRef]
31. Duffy, G.; Lynch, O.A.; Cagney, C. Tracking emerging zoonotic pathogens from farm to fork. *Meat Sci.* **2008**, *78*, 34–42. [CrossRef] [PubMed]
32. Wassenaar, T.M. Following an imaginary Campylobacter population from farm to fork and beyond: A bacterial perspective. *Lett. Appl. Microbiol.* **2011**, *53*, 253–263. [CrossRef] [PubMed]
33. Fozouni, P.; Son, S.; Díaz de León Derby, M.; Knott, G.J.; Gray, C.N.; D'Ambrosio, M.V.; Zhao, C.; Switz, N.A.; Kumar, G.R.; Stephens, S.I.; et al. Direct detection of SARS-CoV-2 using CRISPR-Cas13a and a mobile phone. *medRxiv* **2020**. [CrossRef]
34. Bhunia, A.K.; Taitt, C.R.; Kim, M.S. *High Throughput Screening for Food Safety Assessment*; Woodhead Publishing: Cambridge, UK, 2015. [CrossRef]
35. Varshney, M.; Li, Y. Interdigitated array microelectrode based impedance biosensor coupled with magnetic nanoparticle–antibody conjugates for detection of *Escherichia coli* O157:H7 in food samples. *Biosens. Bioelectron.* **2007**, *22*, 2408–2414. [CrossRef]
36. Daniels, J.S.; Pourmand, N. Label-Free Impedance Biosensors: Opportunities and Challenges. *Electroanalysis* **2007**, *19*, 1239–1257. [CrossRef]
37. Jenkins, D.M.; Lee, B.E.; Jun, S.; Reyes-De-Corcuera, J.; McLamore, E.S. ABE-Stat, a Fully Open-Source and Versatile Wireless Potentiostat Project Including Electrochemical Impedance Spectroscopy. *J. Electrochem. Soc.* **2019**, *166*, B3056–B3065. [CrossRef]
38. Wang, X.; Feng, Y.; Dong, P.; Huang, J. A Mini Review on Carbon Quantum Dots: Preparation, Properties, and Electrocatalytic Application. *Front. Chem.* **2019**, *7*, 671. [CrossRef]
39. Xiao, Z.; Kong, L.B.; Ruan, S.; Li, X.; Yu, S.; Li, X.; Jiang, Y.; Yao, Z.; Ye, S.; Wang, C.; et al. Recent development in nanocarbon materials for gas sensor applications. *Sens. Actuators B Chem.* **2018**, *274*, 235–267. [CrossRef]
40. Taguchi, M.; Schwalb, N.; Rong, Y.; Vanegas, D.C.; Garland, N.; Tan, M.; Yamaguchi, H.; Claussen, J.C.; McLamore, E.S. pulSED: Pulsed sonoelectrodeposition of fractal nanoplatinum for enhancing amperometric biosensor performance. *Analyst* **2016**, *141*, 3367–3378. [CrossRef]
41. Burrs, S.L.; Vanegas, D.C.; Bhargava, M.; Mechulan, N.; Hendershot, P.; Yamaguchi, H.; Gomes, C.; McLamore, E.S. A comparative study of graphene–hydrogel hybrid bionanocomposites for biosensing. *Analyst* **2015**, *140*, 1466–1476. [CrossRef] [PubMed]
42. Vanegas, D.C.; Taguchi, M.; Chaturvedi, P.; Burrs, S.; Tan, M.; Yamaguchi, H.; McLamore, E.S. A comparative study of carbon–platinum hybrid nanostructure architecture for amperometric biosensing. *Analyst* **2014**, *139*, 660–667. [CrossRef] [PubMed]

43. McLamore, E.S.; Mohanty, S.; Shi, J.; Claussen, J.; Jedlicka, S.S.; Rickus, J.L.; Porterfield, D.M. A self-referencing glutamate biosensor for measuring real time neuronal glutamate flux. *J. Neurosci. Methods* **2010**, *189*, 14–22. [CrossRef] [PubMed]
44. Claussen, J.C.; Hengenius, J.B.; Wickner, M.M.; Fisher, T.S.; Umulis, D.M.; Porterfield, D.M. Effects of Carbon Nanotube-Tethered Nanosphere Density on Amperometric Biosensing: Simulation and Experiment. *J. Phys. Chem. C* **2011**, *115*, 20896–20904. [CrossRef]
45. Claussen, J.C.; Franklin, A.D.; Haque, A.; Porterfield, D.M.; Fisher, T.S. Electrochemical Biosensor of Nanocube-Augmented Carbon Nanotube Networks. *ACS Nano* **2009**, *3*, 37–44. [CrossRef]
46. He, Q.; Das, S.R.; Garland, N.T.; Jing, D.; Hondred, J.A.; Cargill, A.A.; Ding, S.; Karunakaran, C.; Claussen, J.C. Enabling Inkjet Printed Graphene for Ion Selective Electrodes with Postprint Thermal Annealing. *ACS Appl. Mater. Interfaces* **2017**, *9*, 12719–12727. [CrossRef]
47. Taguchi, M.; Ptitsyn, A.; McLamore, E.S.; Claussen, J.C. Nanomaterial-mediated Biosensors for Monitoring Glucose. *J. Diabetes Sci. Technol.* **2014**, *8*, 403–411. [CrossRef]
48. Walcarius, A.; Minteer, S.D.; Wang, J.; Lin, Y.; Merkoçi, A. Nanomaterials for bio-functionalized electrodes: Recent trends. *J. Mater. Chem. B* **2013**, *1*, 4878–4908. [CrossRef]
49. Hills, K.D.; Oliveira, D.A.; Cavallaro, N.D.; Gomes, C.L.; McLamore, E.S. Actuation of chitosan-aptamer nanobrush borders for pathogen sensing. *Analyst* **2018**, *143*, 1650–1661. [CrossRef]
50. Li, X.; ShamsiJazeyi, H.; Pesek, S.L.; Agrawal, A.; Hammouda, B.; Verduzco, R. Thermoresponsive PNIPAAM bottlebrush polymers with tailored side-chain length and end-group structure. *Soft Matter* **2014**, *10*, 2008–2015. [CrossRef]
51. Yin, Z.; Zhang, J.; Jiang, L.-P.; Zhu, J.-J. Thermosensitive Behavior of Poly(N-isopropylacrylamide) and Release of Incorporated Hemoglobin. *J. Phys. Chem. C* **2009**, *113*, 16104–16109. [CrossRef]
52. Leal, D.; De Borggraeve, W.; Encinas, M.V.; Matsuhiro, B.; Müller, R. Preparation and characterization of hydrogels based on homopolymeric fractions of sodium alginate and PNIPAAm. *Carbohydr. Polym.* **2013**, *92*, 157–166. [CrossRef] [PubMed]
53. Liu, Y.; Meng, S.; Mu, L.; Jin, G.; Zhong, W.; Kong, J. Novel renewable immunosensors based on temperature-sensitive PNIPAAm bioconjugates. *Biosens. Bioelectron.* **2008**, *24*, 710–715. [CrossRef] [PubMed]
54. Zhao, X.; Liu, Y.; Lu, J.; Zhou, J.; Li, J. Temperature-Responsive Polymer/Carbon Nanotube Hybrids: Smart Conductive Nanocomposite Films for Modulating the Bioelectrocatalysis of NADH. *Chem. A Eur. J.* **2012**, *18*, 3687–3694. [CrossRef] [PubMed]
55. Gamella, M.; Campuzano, S.; Parrado, C.; Reviejo, A.J.; Pingarrón, J.M. Microorganisms recognition and quantification by lectin adsorptive affinity impedance. *Talanta* **2009**, *78*, 1303–1309. [CrossRef] [PubMed]
56. Campuzano, S.; Orozco, J.; Kagan, D.; Guix, M.; Gao, W.; Sattayasamitsathit, S.; Claussen, J.C.; Merkoçi, A.; Wang, J. Bacterial Isolation by Lectin-Modified Microengines. *Nano Lett.* **2012**, *12*, 396–401. [CrossRef]
57. Vanegas, D.C.; Clark, G.; Cannon, A.E.; Roux, S.; Chaturvedi, P.; McLamore, E.S. A self-referencing biosensor for real-time monitoring of physiological ATP transport in plant systems. *Biosens. Bioelectron.* **2015**, *74*, 37–44. [CrossRef]
58. Vanegas, D.C.; Patiño, L.; Mendez, C.; Oliveira, D.A.D.; Torres, A.M.; Gomes, C.L.; McLamore, E.S. Laser Scribed Graphene Biosensor for Detection of Biogenic Amines in Food Samples Using Locally Sourced Materials. *Biosensors* **2018**, *8*, 42. [CrossRef]
59. Burrs, S.L.; Bhargava, M.; Sidhu, R.; Kiernan-Lewis, J.; Gomes, C.; Claussen, J.C.; McLamore, E.S. A paper based graphene-nanocauliflower hybrid composite for point of care biosensing. *Biosens. Bioelectron.* **2016**, *85*, 479–487. [CrossRef]
60. Soares, R.R.A.; Hjort, R.G.; Pola, C.C.; Parate, K.; Reis, E.L.; Soares, N.F.F.; McLamore, E.S.; Claussen, J.C.; Gomes, C.L. Laser-Induced Graphene Electrochemical Immunosensors for Rapid and Label-Free Monitoring of Salmonella enterica in Chicken Broth. *ACS Sens.* **2020**, *5*, 1900–1911. [CrossRef]
61. Sidhu, R.K.; Cavallaro, N.D.; Pola, C.C.; Danyluk, M.D.; McLamore, E.S.; Gomes, C.L. Planar Interdigitated Aptasensor for Flow-Through Detection of Listeria spp. in Hydroponic Lettuce Growth Media. *Sensors* **2020**, *20*, 5773. [CrossRef] [PubMed]
62. Marhaenanto, B.; Soni, P.; Salakhe, V.M. Development of an internet-based greenhouse control system. *Int. Agric. Eng. J.* **2013**, *22*, 72–83.
63. McLamore, E.S.; Jaroch, D.; Chatni, M.R.; Porterfield, D.M. Self-referencing optrodes for measuring spatially resolved, real-time metabolic oxygen flux in plant systems. *Planta* **2010**, *232*, 1087–1099. [CrossRef] [PubMed]
64. Dean, D.A.; Ramanathan, T.; Machado, D.; Sundararajan, R. Electrical Impedance Spectroscopy Study of Biological Tissues. *J. Electrost.* **2008**, *66*, 165–177. [CrossRef]
65. Ding, Y.; Nawroth, J.C.; McFall-Ngai, M.J.; Kanso, E. Mixing and transport by ciliary carpets: A numerical study. *J. Fluid Mech.* **2014**, *743*, 124–140. [CrossRef]
66. Tsumori, F.; Saijou, A.; Osada, T.; Miura, H. Development of actuation system for artificial cilia with magnetic elastomer. *Jpn. J. Appl. Phys.* **2015**, *54*. [CrossRef]
67. Hanasoge, S.; Hesketh, P.J.; Alexeev, A. Metachronal motion of artificial magnetic cilia. *Soft Matter* **2018**, *14*, 3689–3693. [CrossRef]
68. Reeke, G.N.; Becker, J.W.; Edelman, G.M. The covalent and three-dimensional structure of concanavalin A. IV. Atomic coordinates, hydrogen bonding, and quaternary structure. *J. Biol. Chem.* **1975**, *250*, 1525–1547.
69. Castillo-Torres, K.Y.; Arnold, D.P.; McLamore, E.S. Rapid isolation of *Escherichia coli* from water samples using magnetic microdiscs. *Sens. Actuators B Chem.* **2019**, *291*, 58–66. [CrossRef]
70. Danielsen, E.M.; Hansen, G.H. Lipid rafts in epithelial brush borders: Atypical membrane microdomains with specialized functions. *Biochim. Biophys. Acta Biomembr.* **2003**, *1617*, 1–9. [CrossRef]
71. Samuel, G.; Reeves, P. Biosynthesis of O-antigens: Genes and pathways involved in nucleotide sugar precursor synthesis and O-antigen assembly. *Carbohydr. Res.* **2003**, *338*, 2503–2519. [CrossRef] [PubMed]

72. Singh, A.K.; Harrison, S.H.; Schoeniger, J.S. Gangliosides as Receptors for Biological Toxins: Development of Sensitive Fluoroimmunoassays Using Ganglioside-Bearing Liposomes. *Anal. Chem.* **2000**, *72*, 6019–6024. [CrossRef] [PubMed]
73. Gibson, D.L.; White, A.P.; Snyder, S.D.; Martin, S.; Heiss, C.; Azadi, P.; Surette, M.; Kay, W.W. Salmonella produces an O-antigen capsule regulated by AgfD and important for environmental persistence. *J. Bacteriol.* **2006**, *188*, 7722–7730. [CrossRef] [PubMed
74. Miura, Y.; Hoshino, Y.; Seto, H. Glycopolymer Nanobiotechnology. *Chem. Rev.* **2016**, *116*, 1673–1692. [CrossRef]
75. McLellan, J.S.; Chen, M.; Joyce, M.G.; Sastry, M.; Stewart-Jones, G.B.E.; Yang, Y.; Zhang, B.; Chen, L.; Srivatsan, S.; Zheng, A.; et al. Structure-Based Design of a Fusion Glycoprotein Vaccine for Respiratory Syncytial Virus. *Science* **2013**, *342*, 592–598. [CrossRef
76. Hsieh, C.-L.; Goldsmith, J.A.; Schaub, J.M.; DiVenere, A.M.; Kuo, H.-C.; Javanmardi, K.; Le, K.C.; Wrapp, D.; Lee, A.G.; Liu, Y.; et al. Structure-based design of prefusion-stabilized SARS-CoV-2 spikes. *Science* **2020**, *369*, 1501–1505. [CrossRef]
77. Radke, S.M.; Alocilja, E.C. A high density microelectrode array biosensor for detection of *E. coli* O157:H7. *Biosens. Bioelectron.* **2005**, *20*, 1662–1667. [CrossRef]
78. Joung, C.-K.; Kim, H.-N.; Lim, M.-C.; Jeon, T.-J.; Kim, H.-Y.; Kim, Y.-R. A nanoporous membrane-based impedimetric immunosensor for label-free detection of pathogenic bacteria in whole milk. *Biosens. Bioelectron.* **2013**, *44*, 210–215. [CrossRef]
79. Chowdhury, A.; Raut, S.A.; Narman, H.S. DA-DRLS: Drift adaptive deep reinforcement learning based scheduling for IoT resource management. *J. Netw. Comput. Appl.* **2019**, *138*, 51–65. [CrossRef]
80. Chowdhury, A.D.; De, A.; Chaudhuri, C.R.; Bandyopadhyay, K.; Sen, P. Label free polyaniline based impedimetric biosensor for detection of *E. coli* O157:H7 Bacteria. *Sens. Actuators B Chem.* **2012**, *171*, 916–923. [CrossRef]
81. Jantra, J.; Kanatharana, P.; Asawatreratanakul, P.; Hedström, M.; Mattiasson, B.; Thavarungkul, P. Real-time label-free affinity biosensors for enumeration of total bacteria based on immobilized concanavalin A. *J. Environ. Sci. Health Part A* **2011**, *46* 1450–1460. [CrossRef] [PubMed]
82. Li, Y.; Cheng, P.; Gong, J.; Fang, L.; Deng, J.; Liang, W.; Zheng, J. Amperometric immunosensor for the detection of *Escherichia coli* O157:H7 in food specimens. *Anal. Biochem.* **2012**, *421*, 227–233. [CrossRef] [PubMed]
83. Lu, Q.; Lin, H.; Ge, S.; Luo, S.; Cai, Q.; Grimes, C.A. Wireless, Remote-Query, and High Sensitivity *Escherichia coli* O157:H7 Biosensor Based on the Recognition Action of Concanavalin A. *Anal. Chem.* **2009**, *81*, 5846–5850. [CrossRef] [PubMed]
84. Aloul, F.; Zahidi, S.; El-Hajj, W. Two factor authentication using mobile phones. In Proceedings of the 2009 IEEE/ACS International Conference on Computer Systems and Applications, Rabat, Morocco, 10–13 May 2009; pp. 641–644.
85. Fielding, R.T. Architectural Styles and the Design of Network-Based Software Architectures. Ph.D. Thesis, University of California Irvine, CA, USA, 2000.
86. Surbatovich, M.; Aljuraidan, J.; Bauer, L.; Das, A.; Jia, L. Some Recipes Can Do More Than Spoil Your Appetite: Analyzing the Security and Privacy Risks of IFTTT Recipes. In Proceedings of the 26th International Conference on World Wide Web, Perth Australia, 3–7 May 2017; pp. 1501–1510.
87. TSN. Time-Sensitive Networking (TSN) Task Group. Available online: https://1.ieee802.org/tsn/ (accessed on 24 October 2020).
88. NIST. NIST Cybersecurity for IoT Program. 2020. Available online: https://www.nist.gov/programs-projects/nist-cybersecurity-iot-program (accessed on 23 August 2020).
89. Datta, S.P.A. Emergence of Digital Twins: Is This the March of Reason? 2016. Available online: https://dspace.mit.edu/handle/1721.1/104429 (accessed on 20 October 2020).
90. CISA. UDP-Based Amplification Attacks. Available online: https://us-cert.cisa.gov/ncas/alerts/TA14-017A (accessed on 25 October 2020).
91. Galeano-Brajones, J.; Carmona-Murillo, J.; Valenzuela-Valdés, J.F.; Luna-Valero, F. Detection and Mitigation of DoS and DDoS Attacks in IoT-Based Stateful SDN: An Experimental Approach. *Sensors* **2020**, *20*, 816. [CrossRef] [PubMed]
92. NIST. Mitigating IoT-Based DDoS. Available online: https://www.nccoe.nist.gov/projects/building-blocks/mitigating-iot-based-ddos (accessed on 25 October 2020).
93. Srinidhi, N.N.; Dilip Kumar, S.M.; Venugopal, K.R. Network optimizations in the Internet of Things: A review. *Eng. Sci. Technol. Int. J.* **2019**, *22*, 1–21. [CrossRef]
94. Lee, E.A. Computing needs time. *Commun. ACM* **2009**, *52*, 70–79. [CrossRef]
95. Humphreys, T.E.; Ledvina, B.M.; Psiaki, M.L.; O'Hanlon, B.W.; Kintner, P.M. Assessing the Spoofing Threat: Development of a Portable GPS Civilian Spoofer. In Proceedings of the ION GNSS Conference, Savannah, GA, USA, 16–19 September 2008.
96. King, C.; Chu, J.; Mellinger, A.O. *Emerging Technology Domains Risk Survey*; Carnegie Mellon University (CMU): Pittsburgh, PA, USA, 2015.
97. Wei, X.; Sikdar, B. Impact of GPS Time Spoofing Attacks on Cyber Physical Systems. In Proceedings of the 2019 IEEE International Conference on Industrial Technology (ICIT), Melbourne, Australia, 13–15 February 2019; pp. 1155–1160.
98. Zhao, B.; Chen, Q. Location Spoofing in a Location-Based Game: A Case Study of Pokémon Go. In Proceedings of the International Cartographic Conference, Durban, South Africa, 10–16 August 2003; pp. 21–32.
99. National Institute of Standards and Technology. *Cybersecurity Profile for the Responsible Use of Positioning Navigation and Timing (PNT) Sevices*; NIST: Gaithersburg, MD, USA, 2020.
100. Psiaki, M.L.; Humphreys, T.E. Protecting GPS from Spoofers is Critical to the Future of Navigation. Available online: https://spectrum.ieee.org/telecom/security/protecting-gps-from-spoofers-is-critical-to-the-future-of-navigation (accessed on 25 October 2020).

101. Eldosouky, A.; Ferdowsi, A.; Saad, W. Drones in Distress: A Game-Theoretic Countermeasure for Protecting UAVs against GPS Spoofing. *IEEE Internet Things J.* **2020**, *7*, 2840–2854. [CrossRef]
102. Mallikarjun, B.C.; Kiranmayi, K.J.; Lavanya, N.; Prateeksha, K.H.; Sushmitha, J. Intruder Detection System—A LoRa Based Approach. In Proceedings of the 2020 5th International Conference on Communication and Electronics Systems (ICCES), Coimbatore, India, 10–12 June 2020; pp. 255–258.
103. Microsoft. Azure Active Directory Authentication Documentation. Available online: https://docs.microsoft.com/en-us/azure/active-directory/authentication/ (accessed on 24 October 2020).
104. Das, M.L.; Samdaria, N. On the security of SSL/TLS-enabled applications. *Appl. Comput. Inform.* **2014**, *10*, 68–81. [CrossRef]
105. MIT. Certificates. Available online: https://ist.mit.edu/certificates (accessed on 25 October 2020).
106. Sarma, S.; Brock, D.; Engels, D. Radio frequency identification and the electronic product code. *IEEE Micro* **2001**, *21*, 50–54. [CrossRef]
107. Sarma, S.E.; Weis, S.A.; Engels, D.W. *RFID Systems and Security and Privacy Implications*; Springer: Berlin/Heidelberg, Germany, 2002; pp. 454–469.
108. Shabandri, B.; Maheshwari, P. Enhancing IoT Security and Privacy Using Distributed Ledgers with IOTA and the Tangle. In Proceedings of the 2019 6th International Conference on Signal Processing and Integrated Networks (SPIN), Noida, India, 7–8 March 2019; pp. 1069–1075.
109. Djaja, F. Send IoT Data to the IOTA Tangle with SAP HANA XSA and Analytics Cloud. Available online: https://blogs.sap.com/2019/10/08/send-iot-data-to-the-iota-tangle-with-sap-hana-xsa-and-analytics-cloud/ (accessed on 24 October 2020).

actuators

MDPI

Article

Wireless Walking Paper Robot Driven by Magnetic Polymer Actuator

Han-Sol Lee [1,2], Yong-Uk Jeon [1], In-Seong Lee [1], Jin-Yong Jeong [1], Manh Cuong Hoang [1,2], Ayoung Hong [3], Eunpyo Choi [1,2], Jong-Oh Park [1,2] and Chang-Sei Kim [1,2,*]

1 School of Mechanical Engineering, Chonnam National University, Gwangju 61186, Korea; hansol3607@gmail.com (H.-S.L.); jyu9515@naver.com (Y.-U.J.); dlstjd8044@gmail.com (I.-S.L.); s001wlsdyd@naver.com (J.-Y.J.); hmcuong.hust@gmail.com (M.C.H.); eunpyochoi@jnu.ac.kr (E.C.); jop@kimiro.re.kr (J.-O.P.)

2 Korea Institute of Medical Microrobotics, Gwangju 61011, Korea

3 Robotics Engineering, Chonnam National University, Gwangju 61186, Korea; ahong@jnu.ac.kr

* Correspondence: ckim@jnu.ac.kr; Tel.: +82-62-530-5260

Received: 23 September 2020; Accepted: 27 October 2020; Published: 30 October 2020

Abstract: Untethered small-scale soft robots have been widely researched because they can be employed to perform wireless procedures via natural orifices in the human body, or other minimally invasive operations. Nevertheless, achieving untethered robotic motion remains challenging owing to the lack of an effective wireless actuation mechanism. To overcome this limitation, we propose a magnetically actuated walking soft robot based on paper and a chained magnetic-microparticle-embedded polymer actuator. The magnetic polymer actuator was prepared by combining Fe_3O_4 magnetic particles (MPs, diameter of ~50 nm) and silicon that are affected by a magnetic field; thereafter, the magnetic properties were quantified to achieve proper force and optimized according to the mass ratio, viscosity, and rotational speed of a spin coater. The fabricated polymer was utilized as a soft robot actuator that can be controlled using an external magnetic field, and paper was employed to construct the robot body with legs to achieve walking motion. To confirm the feasibility of the designed robot, the operating capability of the robot was analyzed through finite element simulation, and a walking experiment was conducted using electromagnetic actuation. The soft robot could be moved by varying the magnetic flux density and on–off state, and it demonstrated a maximum moving speed of 0.77 mm/s. Further studies on the proposed soft walking robot may advance the development of small-scale robots with diagnostic and therapeutic functionalities for application in biomedical fields.

Keywords: soft robot; paper robot; magnetic polymer; electromagnetic actuation

1. Introduction

Small-scale soft robots that do not require an internal power source and energy transmission parts to exhibit movements can be applied in a limited space, such as in the case of biomedical research [1,2]. In addition, unlike conventional large-scale robots that employ bulky actuators and motors [3,4], these robots require small-sized actuators; this serves as motivation for the development of soft robots using various composite polymers and materials. Inspired by the movements of animals and insects in nature, several researchers have developed robots by simulating their motion mechanisms; moreover, they have also studied multi-functional robots with limited and specific structures in varied environments. Recent achievements include milli- and micro-scale robots such as swimming robots [5], crawling robots [6,7], and insect-mimicking robots [8,9]. Particularly, actuators that can be controlled using magnetic field [10–12], light [13,14], and heat [15,16] were used to achieve the motion of robots. Furthermore, there are examples of robots whose skeletal structure comprises paper [17–19], shape memory alloys [7,20], or polymers [1].

Paper-based robotics research has been introduced for various applications in soft robotics [21,22]. Because paper is inexpensive and easily accessible, it has the benefit of quick and easy fabrication while prototyping. Moreover, paper demonstrates its elastic and plastic deformation properties when folded or unfolded, which is useful to build up joints and links in a robot body. By attaching an actuator and a sensor including an electric circuit, robotic motion can be accomplished. For example, a paper-based actuator that performs a bending or unfolding motion has been developed [23,24], and it has been applied to achieve a gripping motion [25]. Self-folding could be realized by attaching an additional actuator to the paper structure [18]. Furthermore, a wall-climbing robot that employs the elastic energy of paper to achieve a self-folding motion was developed [26]. There are also soft robots that perform deformations or other motions using paper origami structures [17,27,28].

As demonstrated in previous studies, robots based on paper structures mainly exhibit structural folding and bending motions. This requires the attachment of an actuator to the paper, and an electrical signal is generally applied as a stimulation factor via a microcontroller. Such a wired robot has a disadvantage that it can be substantially affected by environmental factors. Particularly, in small-scale soft robots, which are developed for performing technical operations in isolated spaces such as organs, this would lead to critical defects. Notably, robots with a complex structure, such as the origami structure, include a plurality of actuators for operation; therefore, the advantages of inexpensive and reproducible paper materials cannot be maximized.

In this paper, we present a walking soft robot using paper and magnetic polymer actuators that respond to an external magnetic field. The skeleton of the proposed robot was made of paper which can work as folded joints in the robot body. In addition, a polymer based on a mixture of magnetic particles (MPs) and silicon was applied to the actuator part to impart a kinematic change. The walking mechanism of the proposed robot could be controlled through an electromagnetic actuation (EMA) system [29–31]. The magneto-responsive movement of the magnetic polymer actuator was transferred to the paper skeleton; consequently, the robotic walking motion was varied by changing the on–off states of the input magnetic field. Finally, the walking motion was realized based on the frictional force with the floor and the elastic force generated at the joint. We optimized the robot structure through kinematical and material characterization of the robot; thus, an analysis of robot motion and speed spectrum with magnetic flux density variation was performed, and structural parameters were obtained. The proposed method has a high reproducibility owing to the cost-effectiveness and simplicity of manufacturing a paper structure; moreover, it has the advantage of active motion control inside an isolated space via the EMA remote control system. In addition, the magnetic polymer actuator does not use an electronic plate that includes coils or a permanent magnet; therefore, it can be applied flexibly in varied environments without being adversely affected by problems such as corrosion.

The remainder of this paper is organized as follows: We have elucidated the design of the paper robot, along with its kinematics and the fabrication of the magnetic polymer actuator, in Section 2. In Section 3, the characteristics and evaluation of the magnetic polymer actuator and the paper material are presented. The robotic motion simulation and experimental results are presented in Section 4. Finally, concluding remarks, along with a discussion of avenues for future work based on the proposed method are presented in Section 5.

2. Materials and Methods

2.1. Design of a Walking Soft Robot

Designing a soft robot based on magnetic polymer actuators is a key issue for establishing an operating mechanism. In this section, we explain the structure of the proposed soft robot and its magnetic field-based walking mechanism. The structure of the proposed robot fundamentally consists of eight axes, as presented in Figure 1a; each joint is connected by eight rigid links. Here, magnetic polymer actuators were attached to links 1, 3, and 7. Further, polymers composed of aligned MPs (Section 2.2) were attached to each link so that a torque could be applied to achieve robot motion.

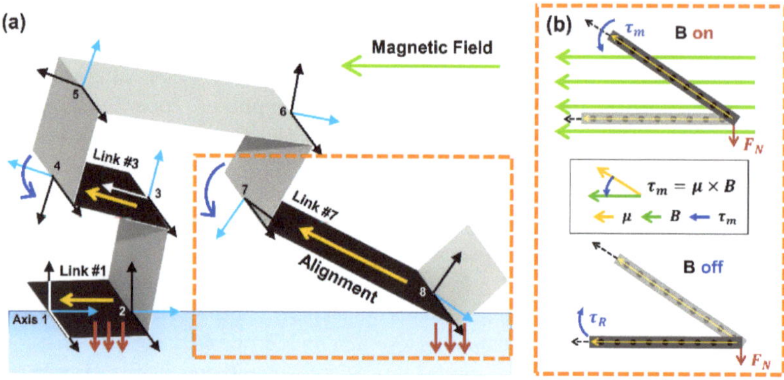

Figure 1. (**a**) Kinematic representation of the proposed walking soft robot; coordinate definition and geometric discretization of the robot. Blue axis means x-coordinates, other axes mean y-, z-coordinates. Black planes are where the magnetic polymers are attached, and red arrows are the normal force of the robot. (**b**) Schematic diagram of the of torque generation mechanism using magnetic polymer and magnetic field alignment. (τ : Torque, μ: Magnetic moment of MPs, B: Magnetic field).

The operating mechanism of the magnetic polymer consists of two motions according to the on-and-off state of the magnetic field, and the operating state of link 7 is described in Figure 1b. Initially, when the magnetic field is not present, only the normal force (F_N) due to gravity is maintained. Under the influence of the magnetic field (B), the polymer rotates and bends around axis 8 by the magnetic moment (μ) caused by the aligned particles and the torque (τ_m) generated through the interaction of the external magnetic field. After that, the bending polymer generates a recovery torque (τ_r) according to the elasticity of the paper constituting the structure, which moves the link to the initial position when the magnetic field is turned off, and at the same time causes a driving force for the walking motion. Thus, a motion mechanism based on the magnetic field, gradient, friction force, and elastic force arising from the material of the robot was established, and the kinematic movement and operation mechanism of each link were analyzed.

Forward kinematics were derived for analyzing the robotic motion based on the orientation and position information of the robot link as a function of each joint angle. To derive the kinematics of the proposed walking robot, we used a Denavit–Hartenberg (D–H) matrix that consisted of four parameters: a_{i-1}, α_{i-1}, d_i, and θ_i, which correspond to the link length, link twist, link offset, and joint angle, respectively. Figure 1 depicts the geometric structure of the soft robot and coordinate frame assignment.

The D–H convention matrix is shown in Equation (1).

$$T_i^{i-1} = \begin{bmatrix} \cos\theta_i & -cos\alpha_{i-1}sin\theta_i & sin\alpha_{i-1}sin\theta_i & a_{i-1}cos\theta_i \\ sin\theta_i & cos\alpha_{i-1}cos\theta_i & -sin\alpha_{i-1}cos\theta_i & a_{i-1}sin\theta_i \\ 0 & sin\alpha_{i-1} & cos\alpha_{i-1} & d_i \\ 0 & 0 & 0 & 1 \end{bmatrix} \quad (1)$$

The proposed robot mechanism is actuated by varying the joint angle for the given length of the link. Therefore, the values of constants α_{i-1} and d_i in matrix T_i^{i-1} correspond to zero, whereas the parameters a_{i-1} for a link length and θ_i for a revolute joint are constant and variable, respectively. The D–H parameters are listed in Table 1. These values were set within the variable range, assuming that force is applied to links 1, 3, and 7 as in Figure 1a; further, they were used to determine the movability of the walking mechanism.

Table 1. Denavit–Hartenberg (D–H) parameters of a walking soft robot.

Link #	Link Length (l_i)	Initial Joint Angle (θ_i)	Joint Angle (θ_j)
1	4	0°	0°
2	5.5	90°	90° < θ_2 < 110°
3	5	75°	75° < θ_3 < 80°
4	5	−80°	−110° < θ_4 < −80°
5	11	−95°	−105° < θ_5 < −95°
6	4.5	−100°	−100° < θ_6 < −75°
7	11.5	80°	85° < θ_7 < 100°
8	3	75°	75°

Using (1) and the D–H parameters, the T_i^{i-1} matrices of each joint could be derived. Moreover, we can obtain the individual transformation matrices $T_1^0 \ldots T_8^7$ and the global transformation matrices $T_{L_n}^0$ via (2).

$$T_{L_n}^R = T_{L_1}^R\, T_{L_2}^{L_1}\, T_{L_3}^{L_2} \cdots T_{L_n}^{L_{n-1}} \quad (2)$$

Furthermore, by using the $T_{L_n}^0$ matrices, the position of links that comprise the robot could be determined. We examined the walking motion according to the change in the link position using the derived kinematic model. Consequently, the simulated motions are depicted in Figure 2, with respect to the joint angle varying. Position 1 represents the applied initial condition, and motions 2 and 3 represent the cases of applying the medium and maximum deflection states, respectively. This shows that, if a torque that is sufficient to cause a change in the parameters listed in Table 1 can be applied to each link, a contraction motion could be realized for the elastic propulsion of a paper robot. Moreover, in position 3, in the absence of magnetic field, the contracted soft robot moves forward owing to the torque generated by the elastic force of the paper and friction.

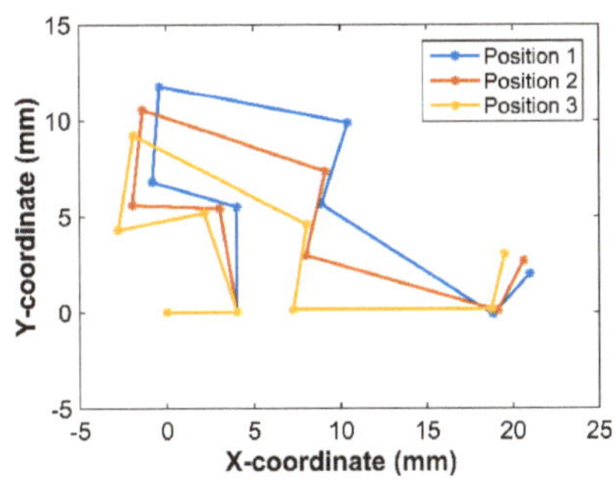

Figure 2. Forward kinematics simulation representing the link position with joint angle variation.

The initial design confirmed through forward kinematics is to examine the operability of the robot. As mentioned above, the designed robot realizes motion by the interaction of forces transmitted to the link, and structurally, the length of the link acts as the most fatal factor to the motion. Intuitively, links 1 to 3 are parts for promoting the working motion, 6 to 8 are parts to support the motion that will occur during movement, and link 5 is a part that organically connects these two roles. In detail, the length of the 3rd link (position 3 in Figure 2) should be long enough to induce torque by tilting the second link forward, so that when the magnetic field is removed, links 1 and 2 are pushed forward. The length of link 7 should be designed to provide the torque and friction force that will cause the robot to take the squat motion. If links 1, 3, and 7 are too short and do not bend properly in the gravitational direction, a problem may arise in generating a force interaction for propulsion. Therefore, it was considered that the designed robot maintains the complementary mechanical form of the link for taking motion, and the process of detailed calibration of this through simulation and experiment was carried out in Section 4.1.

2.2. Magnetic Polymer Actuator Materials

The skeleton structure and working components of the soft robot were constructed using paper and the aforementioned polymer actuator. Here, because the magnetic polymer acts as an actuator that imparts force to the soft robot, the materials contained in the polymer composite should be characterized. For magnetic polymer fabrication, Fe_3O_4 (diameter of ~50 nm) and Ecoflex 00-30 were purchased from Sigma-Aldrich (St. Louis, USA) and Smooth-On Inc. (Macungie, USA), respectively. Fe_3O_4 was selected for its biocompatibility because of it has a relatively low toxicity compared to other MPs. A polymer obtained by spin coating Ecoflex 00-30 silicon containing a specific ratio of Fe_3O_4 was prepared, and inspection was performed via scanning electron microscopy (SEM, SU-8000) to confirm its morphology and surface condition. Using an EMA system and vibrating sample magnetometer (VSM, lake shore 7400 series), the reaction and motion test of magnetic polymer actuator was performed.

2.3. Fabrication of Magnetic Polymer Actuator

The magnetic polymer actuator is composed of Fe_3O_4 as the magnetic particle and Ecoflex 00-30 silicon. As depicted in Figure 3, to obtain the polymer solution, fabrication proceeded in the following order. Firstly, the silicon subject and the curing agent were slowly stirred in the ratio 1:1 to prevent air bubbles. In the case of using Ecoflex 00-20, it caused lower tensile strength than Ecoflex 00-30 (200 psi), and there was no difference in mixing viscosity (3000 cps), specific gravity (1.07 g/cc), and volume (26 in/lb). Thereafter, Fe_3O_4 was mixed with Ecoflex 00-30 at 20, 30, and 40 wt% to prepare a solution of the polymer and applied onto a slide glass (Figure 3a). The sample was spin-coated at 600 and 800 rpm to obtain the desired thickness (Figure 3b). For magnetization, a sample was placed between permanent magnets (neodymium, 50 × 25 × 10 mm), as shown in Figure 3c. Here, the north pole of one magnet was oriented to face the south pole of the other, and the two magnets were set at a distance of 70 mm from one another. The sample was exposed to a magnetic field for 1 min to form an alignment through the magnetization of the internal MPs. At this time, the magnetic field was formed in the horizontal direction along with the permanent magnet placement, and the direction of MPs' arrangement coincided with the permanent magnet. Finally, the magnetic polymer fabrication was complete when the sample separated from the slide glass after curing (Figure 3d).

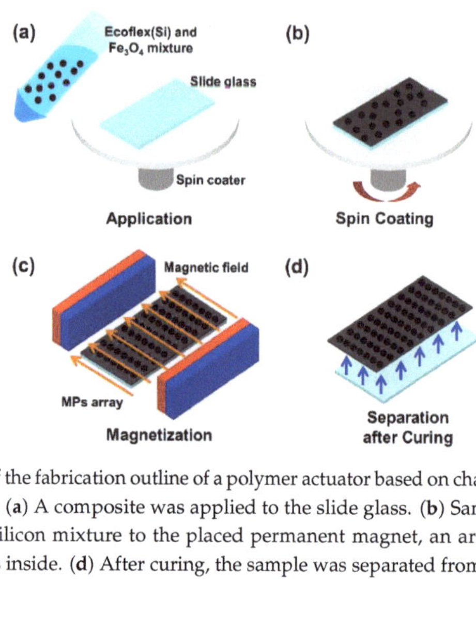

Figure 3. Schematic of the fabrication outline of a polymer actuator based on chained-magnetic-particle-embedded elastomer. (**a**) A composite was applied to the slide glass. (**b**) Samples were spin coated. (**c**) By exposing the silicon mixture to the placed permanent magnet, an array was formed by the magnetization of MPs inside. (**d**) After curing, the sample was separated from the slide glass.

3. Characterization

3.1. Magnetic Polymer Actuator

To evaluate the reactivity of the fabricated magnetic polymer, property tests were conducted in terms of two parameters: (1) the magnetic content of the polymer actuator; (2) the distance from a permanent magnet. Taking this into consideration, the test proceeded by attaching the polymer edge to a fixed stage to react only through bending and compared the displacement of the changing shadow while moving the position of the permanent magnet (Figure 4a). Polymers prepared for characterization consisted of 20, 30, and 40 wt%. Here, if the content of the MPs is increased to 50 wt% or more, the remanence magnetization increases in the fabrication step 3 (Figure 3b), adversely affecting the magnetic particle chain that should be parallel to the magnetic field, and reducing the magnetic field reactivity [2]. By changing the mixing ratio of the Fe_3O_4 solution and the number of spins, the test was repeated five times with every new polymer sample for statistics [32]. As a result, the composition ratio of Fe_3O_4 in the polymer was higher, the deflection according to the magnetic field becomes higher, as provided in Figure 4c,d, where each graph is polynomial curve-fitting data and contains an error of up to 0.5 mm at 600 rpm and up to 1 mm at 800 rpm. From the data, it was confirmed that the deflection at 600 rpm recorded a higher value on average, and that there was a section in which a 40 wt% polymer change rapidly occurred when the permanent magnet was placed within 45 mm at 800 rpm. This is considered to be caused by the decrease in the polymer thickness due to the high rotational speed during the spin coating step. Additionally, for each variable, the cure time of the polymer on average required 1.5 days at 20 wt%, 1.5 days at 30 wt%, and 6 days at 40 wt% (Figure 4b). Based on this, a polymer with a higher reactivity and a shorter cure time with 30 wt% was selected for the polymer actuator.

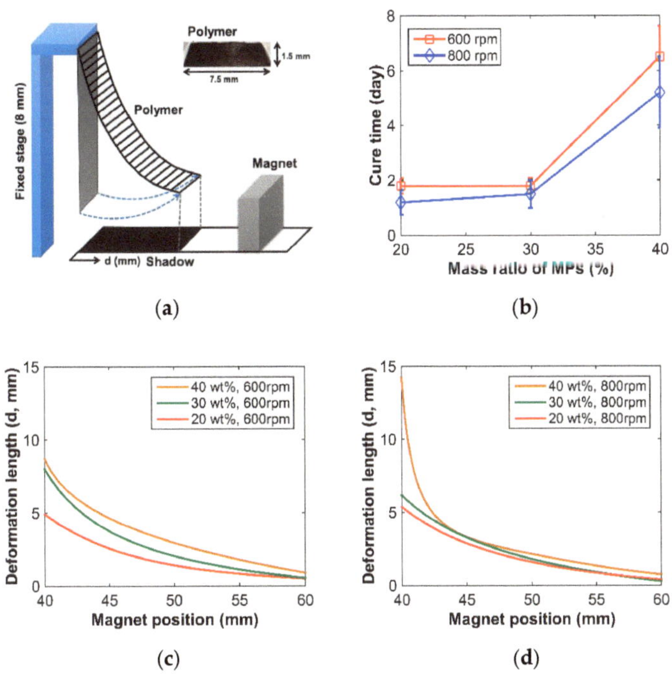

Figure 4. (**a**) Illustration of the test for characterizing the fabricated magnetic polymer. (**b**) Cure time according to the mass ratio of MPs of silicon. (**c**,**d**) Result of the difference in the shadow length with respect to the magnet position when spin coating speeds of (**c**) 600 rpm and (**d**) 800 rpm are applied.

The surface of each polymer was photographed using SEM to precisely observe the polymer morphology and the result of the MPs' alignment with respect to the number of coating revolutions. Figure 5 is an enlarged view of the polymer surface produced by the proposed fabrication method. The left figure is the result of using a 30 wt% solution of Fe_3O_4 and 600 rpm, and the right side is the result of spin coating at 800 rpm. The two pictures above are images of 30 times the polymer surface at each spin coating speed, and the pictures below are 300 times enlarged. Through the SEM images, we judged that there is a uniformity of MPs application through comparison of each spin coating speed and polymer surface state, and we then examined the chain structure in which the difference occurs depending on the applied rotation speed. The wrinkled lines were caused by the shrinkage of the polymer in the curing step, corresponding to Figure 4d. During curing step, both crosslinking of the polymer and evaporation of the solvent caused a shrinkage in the volume of the polymer, further causing the fixed MPs to form a wrinkled structure. The equilibrium wrinkles in the chained samples indicate that the particles in the film are well aligned, and in the unchained samples randomly wrinkled structures are observed. Considering that the directions of the formed wrinkled lines and the left arrow are the same, it was confirmed that the MPs are aligned in the direction of the magnetic field. In addition, the polymer applied with 800 rpm had more wrinkles than that applied with 600 rpm, which was judged by the influence of more evenly distributed MPs at high number of revolutions. However, at the ×30 ratio, it is possible to find a randomly wrinkled section to the right of the section of 600 rpm and 800 rpm. This is the case where MPs are located outside the section where the magnetic field is formed horizontally by the permanent magnet, and the polymer actuator is used as a section in which the uniform alignment is formed. So far, the characterization of the magnetic polymer has been analyzed and appropriate component values have been set.

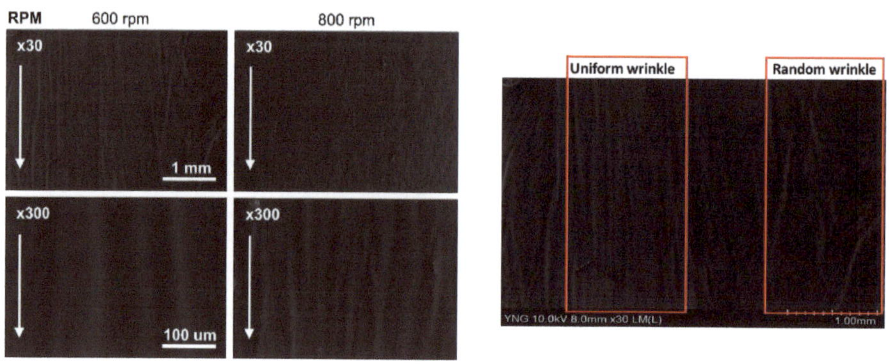

(a) (b)

Figure 5. SEM images of magnetic polymer: (**a**) with 600 (**left**) and 800 (**right**) rpm spin coating speed (Magnetic polymer: 30 wt%). Arrows indicate the magnetic field direction by two permanent magnets, (**b**) uniform wrinkle and random wrinkle comparison.

3.2. Paper Property

It is necessary to present the numerical properties of the paper to specify the characteristics of the paper-based robot and confirm its movability in response to an external magnetic field. The paper density, yield strength and Poisson ratio of the utilized standard office copy paper were 1200 kg/m^3, 33.65 Pa, and 0.27, which are calibrated by simulation. Moreover, the Young's modulus was calculated by using a uniformly loaded cantilever deflection equation.

$$E = wL^4/8I\delta \quad (3)$$

where w, and L are the uniform load and length of the beam, I is an area moment of inertia, and δ is the maximum deflection. For deriving each parameter, we conducted a simple experiment using four paper samples of different sizes. Table 2 includes the parameters of paper samples used for measuring the deflection in each test.

Table 2. Parameters of paper samples.

Paper Sample	Paper Size (mm×mm)	Length (L, mm)	Width (b, mm)	Thickness (h, mm)
1	105 × 148.5	102	148.5	0.13
2	105 × 74.25	102	74.25	0.13
3	52.5 × 74.25	49.5	74.25	0.13
4	52. 5 × 37.125	49.5	37.125	0.13

The test for Young's modulus examination was implemented by the experimental setup depicted in Figure 6a. After using 3 mm of length for fixing, the deflection for the intrinsic weight of each sample was measured. This was measured for five times each through the camera, and the averaged results are presented in Figure 6b along with the calculated value of Young's modulus. When compared with the actual paper Young's modulus value, the calculated value has an error depending on the paper sample size and measurement scale. Even if the deflection of the sample is small, a relatively low Young's modulus may have been calculated by the result derived from the set value, such as the term to the power of 4 of the length. So, we applied the Young's modulus as an average of 577 kPa for subsequent simulations. These characterized values were used to verify the movement of the walking robot.

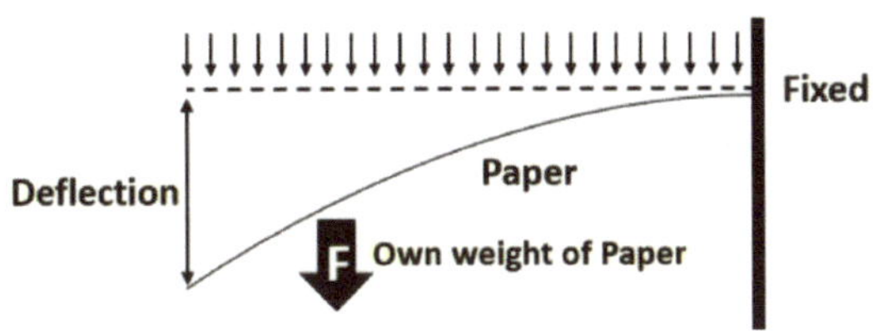

(a)

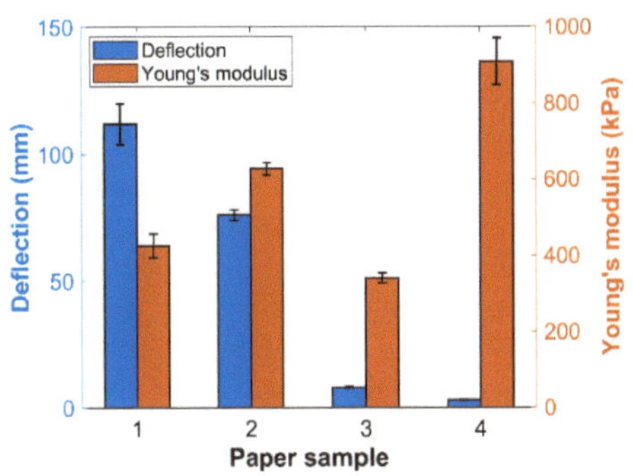

(b)

Figure 6. Characterization of Young's modulus using deflection measurement. (**a**) Schematic images of paper deflection test. (**b**) Deflection measurement value and calculated value of Young's modulus according to the paper sample.

4. Simulation and Experiments

4.1. Robot Motion Simulation

For the verification of whether the paper-based soft robot can execute the motion as depicted in Figure 2, a numerical simulation was performed through the finite element method. The simulation parameters were obtained through the physical properties of the paper constituting the soft robot and the characterization results detailed in Section 3. In which, to minimize the numerical inaccuracy of the alignment of MPs and the physical properties of the polymer actuator, the force due to polymer bending was assumed to be the same as the input force to produce robot joint torque in the presence of a magnetic field. For this, the actual value of the bending force is obtained by the displacement measurement after attaching the polymer to the spring under the on and off states of the magnetic fields. The bending force at the time of applying a magnetic field was derived according to the spring constant. Finally, the force (36.805 N/g) generated per 1 g of unit mass was derived through the weight of the polymer. As with the assumptions mentioned, we applied this force value acting on a polymer actuator in a soft robot model and performed a simulation. The appearance before and after the effect of the magnetic field was analyzed and is presented in Figure 7.

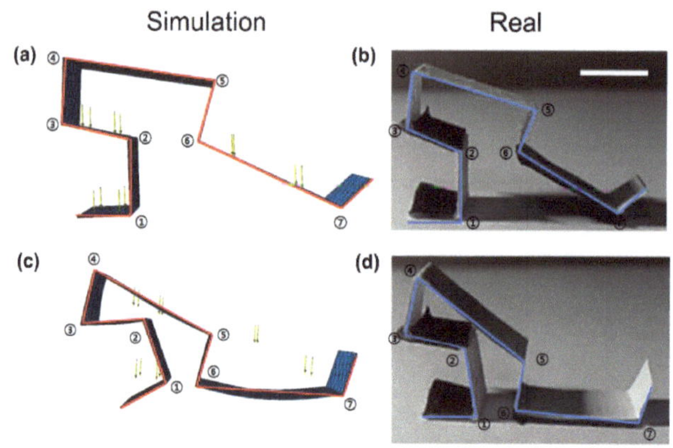

Figure 7. Numerical simulation and actual behavior. (**a**–**d**) The appearance of the soft robot when (**a**,**b**) there is no magnetic field and (**c**,**d**) there is magnetic field of 0.5 T. (Scale bar: 5 mm).

The proposed mechanism used a method of moving forward through the elastic force generated by a paper material with switching from the on-state of the magnetic field to the off-state and the frictional force at the links numbered from one to seven in Figure 7. Based on this, we compared the behavior of the soft robot through simulations and experiments, considering the case of the presence and absence of the magnetic field. In cases where there was no magnetic field, the software robot maintained its initial state as in Figure 7a. Each link maintained a stable state without the influence of external forces other than gravity. Here, when a magnetic field was formed, a bending force was formed in links one, three, and seven, and each link crouched and turned into a shape to move forward. Figure 7c shows the appearance of the soft robot affected by the magnetic field. When a bending force was applied, joints number one to four contracted up to 25° while joint five expanded up to 30°. The robot pulled seven and eight links into a squat motion, and the paper links formed a torque-induced appearance. Thereafter, when the robot switched to the off mode, a clockwise torque occurred, and joint two, three, and four quickly expanded because of the frictional force between links seven and eight, and the floor; thus, the appearance of the robot when it took a step could be predicted.

Subsequently, the reliability of the numerical simulation was verified by comparing each joint change with real robot motions depicted in Figure 7b,d. Prior to comparison, the joints between two adjacent links were numbered and defined as joined in Figure 7, and the numbered angles according to the change in the on–off state of the magnetic field are summarized in Figure 8a. The bars in the graph represent the simulation values, and the error bars represent the error from the experimental values. The soft robot in the off-state was affected only by gravity and exhibited almost the same result as the experimental value; in the on-state, it achieved a coincidence rate of more than 89% on an average. In addition, the simulation results of the soft robots scaled to 1.5 and 2 times also exhibited an average of 95.4% and 86% coincidence when affected by a magnetic field (Figure 8b). Based on this, we predicted the driving mechanism of the model and performed a size standardization and walking experiment of the walking soft robot.

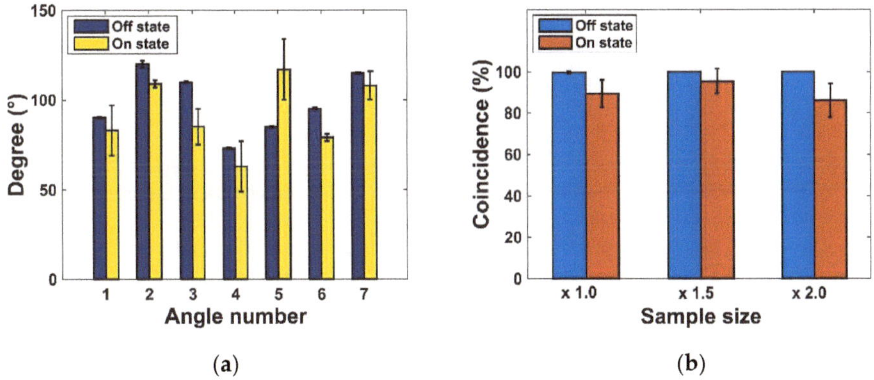

Figure 8. Comparison of angles between joint points of the walking soft robot. (**a**) Joint angle compression result at ×1.0 scale (**b**) Simulation coincidence rate with respect to the sample size.

4.2. Experimental Setup

The EMA system was used for the driving experiment of the walking soft robot. This system consists of a pair of Maxwell coils, a Helmholtz coil, and four rectangular coils, [32,33]. The fabricated soft robot was located in the region of interest for controlling motion. For the behavior of the soft robot, a uniform magnetic field was applied horizontally, and torque (τ) was generated by the elasticity of the paper, $\tau = VM \times B$, where $B = \left[B_x\ B_y\ B_z\right]^T$, V, and $M = \left[m_x\ m_y\ m_z\right]^T$ are the magnetic flux, the volume of the object, and magnetization vector of magnetic device [2]. In addition, it was necessary to form a higher intensity uniform magnetic field in order to achieve the desired operation and speed of various sized robots. For this reason, a VSM (vibrating sample magnetometer, lake shore 7400 series) device was used for generating a larger magnetic field, and the driving efficiency and speed of the soft robot were compared and analyzed based on the change in magnetic field strength of up to 1 T. The on–off control method was operated with a frequency of 2 Hz giving the paper polymer deposits enough time to react to the magnetic field. The experiment was conducted by changing the magnitude of the magnetic field and input signal frequency using the EMA system; soft robots of different sizes were used for the analysis.

4.3. Experimental Results

Figure 9 shows the moving motion over time using a soft robot that was scaled 1.5 times with the highest matching rate in the simulation. We captured every 5 s in the crouched (on) and return (off) states to verify the walking mechanism. When the magnetic field was activated, the soft robot exhibited a shape that curled and contracted while pulling the back leg. Thereafter, along with the removal of the magnetic field, the friction of the back leg suppressed the backwards motion and initiated the forwards motion. Finally, the continuous on–off control of the magnetic field realized the positional movement.

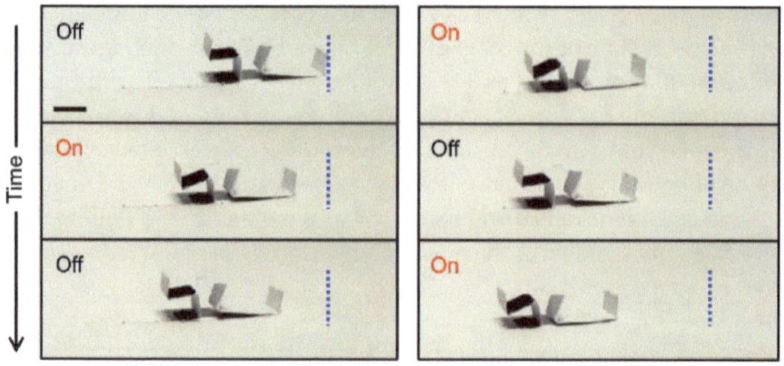

Figure 9. Snapshots of walking soft robot motion. A photograph of the robot's walking motion experiment taken in chronological order. (Blue line: start line, scale bar: 5 mm).

We measured the moving distance and the time taken by the soft robot and accurately calculated the walking velocity according to the structural scale size and magnetic flux density. In order to increase the magnitude of magnetic flux density, experiments were conducted using a VSM device. A uniform magnetic field of maximum 1 T was applied under the same conditions as before. The moving distance was equally applied up to 3 cm, and the elapsed time was measured 10 times for each case. Under the assumption that the walking velocity had a constant value, the velocity in each case was calculated by dividing the travel distance by time. Figure 10 shows the averaged velocity calculated by the experiment according to the abovementioned two variables. The higher flux density increased the velocity, whereas the larger size decreased the velocity. With a magnetic field of 0.649 T and a soft robot with a scale of 1.25 times, a maximum speed of 77 mm/s was achieved. However, in the magnetic field below 0.5 T, the weight of the soft robot had a fatal effect, and it was judged that the generated torque was insufficient to realize the motions of pushing and pulling the paper link. This can be solved by performing studies that would consider the force in motion and develop structures that effectively transmit energy. A material with better resilience could also be used.

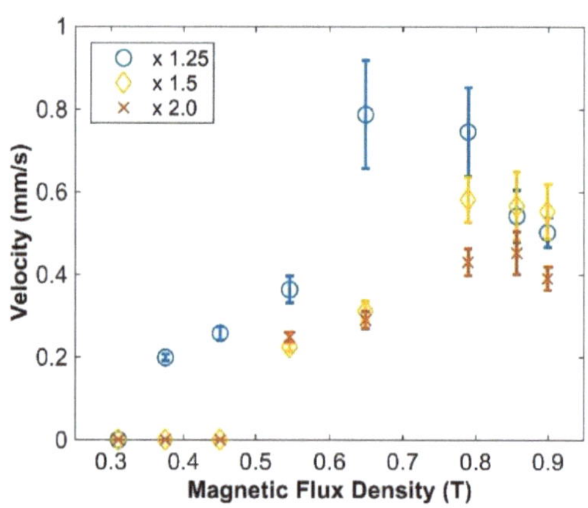

Figure 10. Result of the walking velocity experiment conducted according to the magnetic flux density and varying size of the soft robot.

5. Conclusions

In this study, we proposed and validated a walking soft robot mechanism made of a paper, actuated by a polymer actuator embedded with chained MPs. Magnetic polymers in which MPs are aligned has the property of aligning according to the direction of a uniform magnetic field, and a torque mechanism to generate walking motion was used. The feasibility of the proposed mechanism was examined via robot kinematics and simulation. The proposed soft robot was fabricated in the frame of a paper body and operated by the on and off states of magnetic field control via the EMA system and a VSM for successful motion. The EMA system was useful for controllability such as changes in the direction of the magnetic field, but there was a limit to completely solving the stiffness of the paper with the torque generated by the polymer. Accordingly, a high-intensity magnetic field of up to 1 T was applied using the VSM, and analysis was conducted according to the robot size and walking speed. The devised mechanism was verified, and the performance was evaluated for various sizes of soft robots and a speed control of up to 0.77 mm/s was achieved.

The fabricated magnetic polymer actuator and the paper framework played an important role in realizing a new mechanism for soft walking robots. The proposed soft robot mechanism focuses only on walking motion and its actuation through magnetic polymer actuator; however, to implement a multi-functional small-scale robot, higher degrees of freedom motion and a method for their independent actuation control needs to be conquered. In addition to this, further technical developments including robot joint angle measurements and robot posture sensing, as well as advanced control methods in utilizing active electromagnetic field control are still needed for ensuring the smooth movement of the robot. Although we conducted the robot simulation to verify the motion of the designed robot configuration, the lack of dynamic behavior incorporating the polymer actuator properties needs to be further established to predict the newly designed robot motion before fabrication. Nevertheless, we think that the development of articulated robots through partial actuators suggests a new research direction in the context of recent changes to the polymer model, or in a situation in which soft robots, produced by themselves, are being studied in various fields with improved mechanisms and driving circuits [34,35]. In the future, it is expected that small-scale robots that perform various functionalities in addition to walking will be utilized in various applications such as drug delivery and biopsy in restricted and inaccessible environments within the human body.

Author Contributions: Conceptualization, C.-S.K.; methodology, Y.-U.J., I.-S.L., J.-Y.J., and M.C.H.; software, Y.-U.J., I.-S.L., J.-Y.J., and M.C.H.; validation, H.-S.L. and E.C.; investigation, H.-S.L.; data curation, Y.-U.J., I.-S.L., and J.-Y.J.; writing—original draft preparation, H.-S.L.; writing—review and editing, A.H. and C.-S.K.; visualization, H.-S.L.; supervision, C.-S.K.; funding acquisition, J.-O.P. All authors have read and agreed to the published version of the manuscript.

Funding: This research was supported by a grant of the Korea Health Technology Development R&D Project through the Korea Health Industry Development Institute (KHIDI), funded by the Ministry of Health and Welfare, Republic of Korea (grant number: HI19C0642).

Acknowledgments: We would like to thank anonymous reviewers' valuable comments and helpful discussions to improve the quality of the paper.

Conflicts of Interest: The authors declare no conflict of interest.

References

1. Hu, W.; Lum, G.Z.; Mastrangeli, M.; Sitti, M. Small-scale soft-bodied robot with multimodal locomotion. *Nature* **2018**, *554*, 81–85. [CrossRef] [PubMed]
2. Ijaz, S.; Li, H.; Hoang, M.C.; Kim, C.S.; Bang, D.; Choi, E.; Park, J.O. Magnetically actuated miniature walking soft robot based on chained magnetic microparticles-embedded elastomer. *Sens. Actuators A Phys.* **2020**, *301*, 111707. [CrossRef]
3. Lee, H.D.; Moon, J.I.; Kang, T.H. Design of a series elastic tendon actuator based on gait analysis for a walking assistance exosuit. *Int. J. Control. Autom. Syst.* **2019**, *17*, 2940–2947. [CrossRef]
4. Yazdani, M.; Salarieh, H.; Foumani, M.S. Bio-inspired decentralized architecture for walking of a 5-link biped robot with compliant knee joints. *Int. J. Control. Autom. Syst.* **2018**, *16*, 2935–2947. [CrossRef]
5. Ren, Z.; Hu, W.; Dong, X.; Sitti, M. Multi-functional soft-bodied jellyfish-like swimming. *Nat. Commun.* **2019**, *10*, 1–12. [CrossRef] [PubMed]
6. Shepherd, R.F.; Ilievski, F.; Choi, W.; Morin, S.A.; Stokes, A.A.; Mazzeo, A.D.; Whitesides, G.M. Multigait soft robot. *Proc. Natl. Acad. Sci. USA* **2011**, *108*, 20400–20403. [CrossRef]

7. Yuk, H.; Kim, D.; Lee, H.; Jo, S.; Shin, J.H. Shape memory alloy-based small crawling robots inspired by *C. elegans*. *Bioinspir. Biomim.* **2011**, *6*, 046002. [CrossRef]
8. Wu, Y.; Yim, J.K.; Liang, J.; Shao, Z.; Qi, M.; Zhong, J.; Fearing, R.S. Insect-scale fast moving and ultrarobust soft robot. *Sci. Robot.* **2019**, *4*, eaax1594. [CrossRef]
9. Joyee, E.B.; Pan, Y. Multi-material additive manufacturing of functional soft robot. *Procedia Manuf.* **2019**, *34*, 566–573. [CrossRef]
10. Gray, B.L. A review of magnetic composite polymers applied to microfluidic devices. *J. Electrochem. Soc.* **2014**, *161*, B3173. [CrossRef]
11. Venkiteswaran, V.K.; Samaniego, L.F.P.; Sikorski, J.; Misra, S. Bio-inspired terrestrial motion of magnetic soft millirobots. *IEEE Robot. Autom. Lett.* **2019**, *4*, 1753–1759. [CrossRef]
12. Maeda, K.; Shinoda, H.; Tsumori, F. Miniaturization of worm-type soft robot actuated by magnetic field. *Jpn. J. Appl. Phys.* **2020**, *59*, SIIL04. [CrossRef]
13. Rogóż, M.; Zeng, H.; Xuan, C.; Wiersma, D.S.; Wasylczyk, P. Light-driven soft robot mimics caterpillar locomotion in natural scale. *Adv. Opt. Mater.* **2016**, *4*, 1689–1694. [CrossRef]
14. Pilz da Cunha, M.; Ambergen, S.; Debije, M.G.; Homburg, E.F.; den Toonder, J.M.; Schenning, A.P. A Soft Transporter Robot Fueled by Light. *Adv. Sci.* **2020**, *7*, 1902842. [CrossRef] [PubMed]
15. Cheng, N.; Ishigami, G.; Hawthorne, S.; Chen, H.; Hansen, M.; Telleria, M.; Iagnemma, K. Design and analysis of a soft mobile robot composed of multiple thermally activated joints driven by a single actuator. In Proceedings of the 2010 IEEE International Conference on Robotics and Automation, Anchorage, AK, USA, 3–7 May 2010; pp. 5207–5212.
16. Go, G.; Jin, Z.; Park, J.O.; Park, S. A Thermo-electromagnetically actuated microrobot for the targeted transport of therapeutic agents. *Int. J. Control. Autom. Syst.* **2018**, *16*, 1341–1354. [CrossRef]
17. Shigemune, H.; Maeda, S.; Hara, Y.; Hosoya, N.; Hashimoto, S. Origami robot: A self-folding paper robot with an electrothermal actuator created by printing. *IEEE/ASME Trans. Mechatron.* **2016**, *21*, 2746–2754. [CrossRef]
18. Shigemune, H.; Maeda, S.; Cacucciolo, V.; Iwata, Y.; Iwase, E.; Hashimoto, S.; Sugano, S. Printed paper robot driven by electrostatic actuator. *IEEE Robot. Autom. Lett.* **2017**, *2*, 1001–1007. [CrossRef]
19. Han, D.D.; Zhang, Y.L.; Jiang, H.B.; Xia, H.; Feng, J.; Chen, Q.D.; Sun, H.B. Moisture-responsive graphene paper prepared by self-controlled photoreduction. *Adv. Mater.* **2015**, *27*, 332–338. [CrossRef]
20. Huang, W.; Huang, X.; Majidi, C.; Jawed, M.K. Dynamic simulation of articulated soft robots. *Nat. Commun.* **2020**, *11*, 1–9. [CrossRef]
21. Martinez, R.V.; Fish, C.R.; Chen, X.; Whitesides, G.M. Elastomeric origami: Programmable paper-elastomer composites as pneumatic actuators. *Adv. Funct. Mater.* **2012**, *22*, 1376–1384. [CrossRef]
22. Hamedi, M.M.; Campbell, V.E.; Rothemund, P.; Güder, F.; Christodouleas, D.C.; Bloch, J.F.; Whitesides, G.M. Electrically activated paper actuators. *Adv. Funct. Mater.* **2016**, *26*, 2446–2453. [CrossRef]
23. Amjadi, M.; Sitti, M. High-performance multiresponsive paper actuators. *ACS Nano* **2016**, *10*, 10202–10210. [CrossRef] [PubMed]
24. Wang, G.; Cheng, T.; Do, Y.; Yang, H.; Tao, Y.; Gu, J.; Yao, L. Printed paper actuator: A low-cost reversible actuation and sensing method for shape changing interfaces. In Proceedings of the 2018 CHI Conference on Human Factors in Computing Systems, Montreal, QC, Canada, 21–26 April 2018; pp. 1–12.
25. Ryu, J.; Mohammadifar, M.; Tahernia, M.; Chun, H.I.; Gao, Y.; Choi, S. Paper Robotics: Self-Folding, Gripping, and Locomotion. *Adv. Mater. Technol.* **2020**, *5*, 1901054. [CrossRef]
26. Wu, Q.; Pradeep, V.; Liu, X. A paper-based wall-climbing robot enabled by electrostatic adhesion. In Proceedings of the 2018 IEEE International Conference on Soft Robotics (RoboSoft), Livorno, Italy, 24–28 April 2018; pp. 315–320.
27. Shigemune, H.; Maeda, S.; Hara, Y.; Koike, U.; Hashimoto, S. Kirigami robot: Making paper robot using desktop cutting plotter and inkjet printer. In Proceedings of the 2015 IEEE/RSJ International Conference on Intelligent Robots and Systems (IROS), Hamburg, Germany, 28 September–2 October 2015; pp. 1091–1096.
28. Li, J.; Godaba, H.; Zhang, Z.Q.; Foo, C.C.; Zhu, J. A soft active origami robot. *Extrem. Mech. Lett.* **2018**, *24*, 30–37. [CrossRef]
29. Nguyen, K.T.; Hoang, M.C.; Go, G.; Kang, B.; Choi, E.; Park, J.O.; Kim, C.S. Regularization-based independent control of an external electromagnetic actuator to avoid singularity in the spatial manipulation of a microrobot. *Control Eng. Pract.* **2020**, *97*, 104340. [CrossRef]

30. Hoang, M.C.; Nguyen, K.T.; Le, V.H.; Kim, J.; Choi, E.; Kang, B.; Park, J.O.; Kim, C.S. Independent Electromagnetic Field Control for Practical Approach to Actively Locomotive Wireless Capsule Endoscope. *IEEE Trans. Syst. MAN, Cybern. Syst.* **2019**, *99*, 1–13. [CrossRef]
31. Kim, J.; Nguyen, P.B.; Kang, B.; Choi, E.; Park, J.O.; Kim, C.S. A novel tip-positioning control of a magnetically steerable guidewire in sharply curved blood vessel for percutaneous coronary intervention. *Int. J. Control. Autom. Syst.* **2019**, *17*, 2069–2082. [CrossRef]
32. Liu, J.A.C.; Gillen, J.H.; Mishra, S.R.; Evans, B.A.; Tracy, J.B. Photothermally and magnetically controlled reconfiguration of polymer composites for soft robotics. *Sci. Adv.* **2019**, *5*, eaaw2897. [CrossRef]
33. Hoang, M.C.; Le, V.H.; Kim, J.; Choi, E.; Kang, B.; Park, J.O.; Kim, C.S. Intestinal Tattooing Mechanism Integrated with Active Wireless Capsule Endoscope. In Proceedings of the 2018 7th IEEE International Conference on Biomedical Robotics and Biomechatronics (Biorob), Enschede, The Netherlands, 26–29 August 2018; pp. 1254–1259.
34. Wang, X.; Mao, G.; Ge, J.; Drack, M.; Bermúdez, G.S.C.; Wirthl, D.; Lling, R.; Kosub, T.; Bischoff, L.; Fassbender, J. Untethered and ultrafast soft-bodied robots. *Commun. Mater.* **2020**, *1*, 1–10. [CrossRef]
35. Zhao, J.; Zhang, J.; McCoul, D.; Hao, Z.; Wang, S.; Wang, X.; Sun, L. Soft and fast hopping–running robot with speed of six times its body length per second. *Soft Robot.* **2019**, *6*, 713–721. [CrossRef]

actuators

Article

Presentation, Modeling and Experiments of an Electrostatic Actuator Based Catom for Programmable Matter

Romain Catry [1], Abdenbi Mohand-Ousaid [1], Micky Rakotondrabe [2,*] and Philippe Lutz [1]

1 Department of Automatic Control and Micro-Mechatronic Systems, FEMTO-ST Institute, Université Bourgogne Franche-Comté, CNRS, 24 rue Savary, 25000 Besançon, France; romain.catry@femto-st.fr (R.C.); abdenbi.mohand@femto-st.fr (A.M.-O.); plutz@femto-st.fr (P.L.)
2 Laboratoire Génie de Production, National School of Engineering in Tarbes (ENIT-INPT), University of Toulouse, 47 Avenue d'Azereix, 65000 Tarbes, France
* Correspondence: mrakoton@enit.fr

Received: 9 May 2020; Accepted: 29 May 2020; Published: 4 June 2020

Abstract: Nowadays, the concept of programmable matter paves the way for promising applications such as reshaping an object to test different configurations, modeling or rapid prototyping. Based on elementary modules, such matter can be arranged and disassembled easily according to the needs of the designers. Several solutions have been proposed to implement this concept. Most of them are based on modular self-reconfigurable robotics (SMR) that can work together and move relatively to one another in order to change their configuration. Achieving such behavior requires to solve some technological challenges in particular module's geometry and actuation. In this paper, we build and develop a proof of concept for a catom based on electrostatic actuation. The modeling and analysis of the actuator functioning as catom is given after a comparison of various possible actuation. Simulations as well as experiments validations are afterwards carried out to confirm and demonstrate the efficiency of electrostatic actuation to achieve latching capabilities of the proposed catom.

Keywords: electrostatic actuation; programmable matter; self-reconfigurable robotics

1. Introduction

1.1. Programmable Matter

Programmable matter is a term that was first defined by Toffoli and Margolus [1] to refer to an ensemble of fine grained computing elements arranged in space. It is a matter that can change its properties such as shape, optical properties, color, conductivity, density, and so forth. on demand from a user, from sensor feeds or from software control. A best conceptual illustration of this matter is the fictional T1000 liquid-metal robot from the James Cameron film-Terminator 2: Judgment day [2]. The robot's body, composed of billions of microscopic modules, has the capacity to reconstitute itself after an injury.

One advantage of programmable matter is its morphability which could allow to send or copy objects as easily as digital documents. For example, reshaping an object to test different configurations allows for cheap and rapid prototyping for car manufacturers or plane designers. As described in Figure 1, the modules composing the programmable matter (on the left) can be rearranged to form a car model (on the right). Once the model is validated by designers, they can be disassembled for another use. In an industrial world where time, money and recycling are feature keys, such morphability is a very interesting concept.

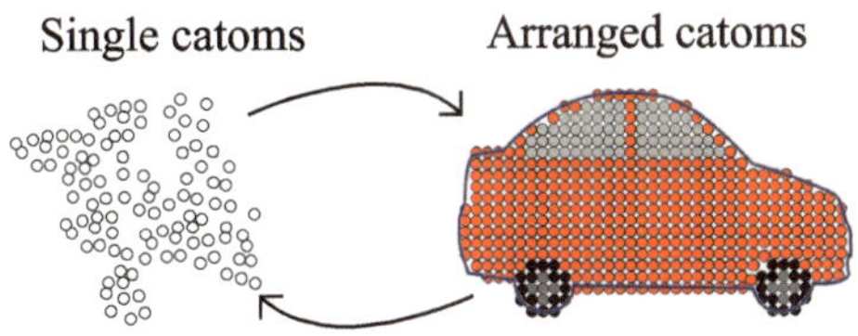

Figure 1. Illustration of programmable matter used for car prototyping unit.

From physical point of view, implementing programmable matter concept is challenging and remains an active research area. Several solutions have been proposed in the literature. Almost all of them are based on modular self-reconfigurable robotics (MSR) which has the advantage of including a computation part. These solutions are composed of individual modules (one or several kind) that have to satisfy the three following properties—Latching, Locomotion, and Communication. Then an algorithm has to overview all the catoms to allow them to arrange their configuration and to reach the desired shape. This paper will study and design one catom that with a focus on the first property: the latching capabilities. In the rest of this introduction-section, we first give a state of the art of programmable matter in MSR approach. Then we compare the different actuation technologies from which we will justify the use of electrostatic actuators in our case. Finally, we present the contribution of the paper.

1.2. State of the Art

In the field of programmable matter and self-reconfigurable robotic systems, there are a lot of works in the literature. Toffoli and Margolus [1] introduced the concept of programmable matter. Later on, roboticists started from this concept in order to go towards physical realization. Fukuda et al. [3] developed the CEBOT (Cellular Robot) and set a base for most of the following works. Many prototypes have been created since then, in a large scope of size, shapes, in both 2D and 3D configurations. While most of the existing systems exploit a lattice structure to lessen the complexity of self-reconfiguration, there also exist chain type systems which do not rely on a lattice. However, these two types do not negate each other and some existing systems are even based on both at the same time such as the M-TRAN (Self-reconfigurable modular robot) [4].

An MSR module has generally to be fast when in movement and has to exhibit a strong latching capability when in rest, and if possible exhibits energy efficiency property. The M-block modules in Reference [5] offer an answer to these requirements by using cubic robots shape able to move thanks to inertial motors and to latch to each other using permanent magnets. They are one of the most advanced modules on the market that can claim to be programmable matter. Their permanent magnets offer strong latching capability, which is very useful to oppose gravity and to stack lots of modules at once. Indeed, gravity is one of the biggest challenge to deal with when working in the 3D MSR. In Reference [6], the authors proposed modules based on helium robots such that the gravity is overcomed. On the other hand, Garcia et al. [7] worked with pneumatic energy to latch their modules. While there was no real actuation, their vacuum powered robots have a very strong latching capability. Such latching strength is one of the most important part regarding programmable matter because the modules have to uphold the desired form when all of them are connected. The Molecule [8] is a set of two modules attached each other by a rigid axis. Each pair of modules can then connect with the other pairs and move around. A lot of robots also use mechanical latching for its strong force generation capability beyond the fact that it is a well known principle. A wide variety of approaches based on mechanical latching exists nowadays. For instance 3D-Units [9] and the ATRON (lattice based self-reconfigurable robot) [10] are based on hooks, whilst the Metamorphic modules in Reference [11] are based on key and locks. Even if lattice based-systems are predominant, one can also find chained based systems in the literature, such as the CONRO (Configurable Robots) [12] or the Modular-Expanding modules [13].

The above survey are related to 3D modules. However, the design of a 3D module can also be inspired from their 2D counterpart, in particular in term of actuation principle. The 2D Claytronic cylinder presented in [14] uses an electrostatic actuation which serves as both latching and movement generation. Though not widely used, the advantage of electrostatic actuation is its miniaturization possibility. In [15], an electromagnetic actuation is used for latching. While one module cannot move relatively to another one, the feature was that they can communicate each other through electromagnetic field In Reference [16], the above electromagnetic actuation for latching has been extended to 3D modules. Yim et al. [2] and Ostergaard et al. [17] provide much more thorough review on modula robotics that can be extended to modules for programmable matter. Overall, most of the current technologies regarding programmable matter favor cube-shaped modules since they offer several advantages: large surfaces for easier latching, ease of fabrication, and high stability when stacked/latched. In counterpart, these advantages are at the expense of the difficulty to create relative movement between modules.

1.3. Comparison of Different Actuation Technologies

To compare the different potential technologies used in modules, we propose the following measures:

- force generated by a module (Latching force),
- energy required for latching (Energy),
- miniaturization possibility (Miniaturization),
- ease of integration of the actuation on a module (Integration),
- and possibility to use the same latching actuation for relative movement (Movement).

Table 1 provides the ranking of the different principles of module latching found in the literature: mechanical latching (Mech), permanent magnet latching (P mag), electromagnetic latching (E mag), electrostatic latching (Elec) and pneumatic latching (Pn). As already stated above, the mechanical latching is one of the most present in current technologies of modules because it is well understood, generates a strong force output and is relatively easy to integrate (mostly for cubic modules). In counterpart mechanical latching is not easy to miniaturize and its energy consumption is relatively high. Permanent magnet latching also suffers from miniaturization limitation. Moreover, it is not adapted as actuation for relative movement between modules because of the strong stuck force they produce and that keep the latters latched, and because the fact that they are not really controllable. On the other hand, electromagnetic actuation is controllable versus permanent magnet actuation. However it requires high current. Beyond that, its main limitation is the very weak force when going to miniaturization. As from Table 1, electrostatic actuation could provide the best compromise about the different criteria we impose to design modules for latching perspective. Even though electrostatic actuation produces less force compared to a mechanical latching or a permanent magnet, it has three main advantages. First, it requires small amounts of energy to ensure the latching. Indeed, electrostatic actuation might require high fixed voltage in certain situation but the electrical current is very weak. Second, the scaling effect allows to obtain higher force density generated by electrostatic actuation and is favorable for miniaturization [18]. This is important for the latching and for the module movement. Finally, this actuation mode can be easily disabled by cutting the power.

Table 1. Comparison table.

Labels	Force	Energy	Miniaturization	Integration	Movement
Mech	2	3	3	2	1
P mag	2	1	4	3	4
E mag	5	4	2	4	2
Elec	4 *	2	1	1	3
Pn	1	5	5	5	5

Mech = Mechanical latching, P mag = permanent magnets, E mag = electromagnets, Elec = Electrostatic latching and Pn = pneumatic latching.
* The smaller the system is, the more potent the electrostatic force will be.

1.4. Contributions

Within the national B3PM (Building the Basic Blocks of Programmable Matter [19]) project devoted to rapid prototyping in automotive industry, as illustrated in Figure 1, a concept of programmable matter has been presented in [20]. Upon this previous work, this paper aims to develop a module based on electrostatic actuation. This actuation mode is expected for both latching and displacement of the module. Furthermore, the design is such that it is suitable for miniaturization purpose. The term catom will be used in the sequel to describe the module. This term comes from the combination of the words Claytronics and atom. The term Claytronics is within programmable matter concept, means "Electronic clay" and combines nanotechnologies and computer science. The catom presented in this work has to fulfill the following specifications:

1. connection between several catoms with minimization of void space between them in order to regularly fill a 3D space,
2. large contact and electrodes surfaces between two catoms in order to maximize the efficiency of electrostatic actuation,
3. centimeter scale for the catom size and ease of miniaturization,
4. mass fabrication possibility.

To address our contributions, the paper is organized as follows. Section 2 describes the choice of the catom geometry, the integration of electrostatic actuation and the modeling of the latching functioning. In Section 3, the realization of both catom's structure and planar electrodes is presented. Experimental validation procedure is presented and the obtained results are discussed in Section 4. Last section summarizes the work and provides some perspectives on further developments.

2. Catom's Shape and Modeling of the Actuation

2.1. Quasi-Spherical Catom as a Module

As pointed previously, the cube shaped catom has several advantages but the sharp edges do not allow an ease of movement. To overcome this limitation, a spherical catom would be the theoretical best structure allowing an ease of movement with the minimal amount of energy. This also allows to mimic nature and the atoms that compose matter. However, a spherical catom is not convenient for latching because of the point-point contact between the two spheres. To tackle this limitation, a quasi-spherical shape concept was adopted in Reference [20] within the B3PM project. In this paper we use the quasi-spherical shape for a catom and proposes to design the actuation that will permit to this to latch and to move.

The suggested quasi-spherical catom has a face-centered cubic (FCC) lattice structure as depicted in Figure 2a. In this configuration each catom can have up to twelve neighbors catoms. Each neighbor catom can be locked onto the dedicated latching surface, called connector, of the initial catom by the proposed electrostatic actuation. The other surfaces (non-connectors) serve as rolling surfaces during the displacement. The length of all the surfaces, connectors and non-connectors, are designed such that a catom in movement can always reach another catom latched and connected on the initial catom. Even if this quasi-spherical shape offers several advantages, we propose a modified wuasi-spherical shaper as illustrated in Figure 2b. In fact, the latching force that will be obtained from the electrostatic actuation is proportional to the connector surface, hence the quasi-spherical shape of Figure 2b will provide better taching strength than that of Figure 2a. In the new suggestion, the connectors surfaces for a given volume in order to ensure the required latching force while rolling and passing surfaces (the non-connectors surfaces) based on triangle empty surfaces are still possible. For the rest of the study, a catom of a diameter of 2 cm is considered with connector's surface of 8 mm × 8 mm.

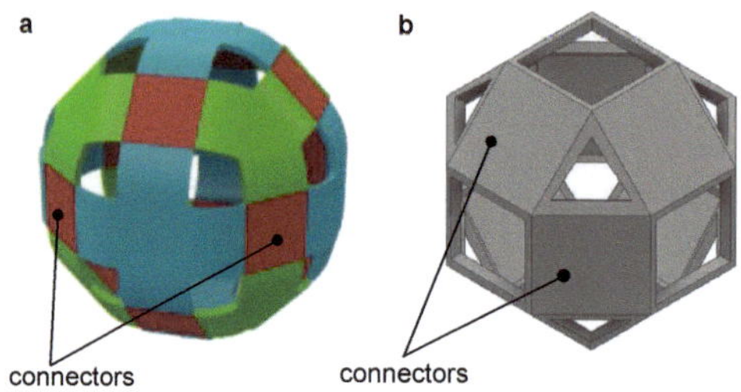

Figure 2. Catom's CAD structure. (**a**) Quasi-spherical catom shape. (**b**) Catom maximizing the surface of connectors.

2.2. Principle of Electrostatic Latching

We present here the principle and model of the electrostatic actuation that will be behind the connectors surfaces and that serve first as latching mechanism. The idea is to distribute electrodes on each connector surface. Hence, when applying an electrical potential to the electrodes on one surface of a catom, and when another potential is applied to the electrodes on the surface of the other catom, the potentials difference will create an attraction force between the two surfaces. Thus, a latching of the two catoms is obtained. The advantages of such electrostatic actuation are—(i) suitable for miniaturization, (ii) easy to implement and (iii) can be used for both latching and displacement actuation and l (iv) low energy consumption. This latter advantage is of particular interest since the theoretical energy consumption is zero in steady state and non zero in transient state, which is very fast. In this paper, we investigate only the latching capability of the catoms. The idea consists in assessing the provided force according to the applied voltage and the geometrical features of the electrodes on each connectors surfaces. In particular, we will study and derive the required force to maintain two catoms attached to each others.

Figure 3 illustrates the principle of the electrostatic actuation where two electrodes surfaces facing each other are electrically charged. Considering this configuration as a parallel capacitor, it is easy to derive the attractive force F between them using the principle of virtual work:

$$\vec{F} = -\vec{\nabla} W_e, \tag{1}$$

where W_e is the stored electrical energy in the capacitor. In the case of parallel capacitor the following expression of the force along the vertical axis can be derived.

$$F = \frac{\varepsilon_0 A}{2(\frac{x_{ins}}{\varepsilon_r} + x_{air})^2} U^2, \tag{2}$$

where ε_0 and ε_r are respectively the dielectric permittivity of the void and the relative permittivity of the insulation material, A is the area of a connector surface, x_{ins} is the width of the insulation, x_{air} is the distance of the air gap separating the two plates and U is the voltage or potentials difference between the two electrodes.

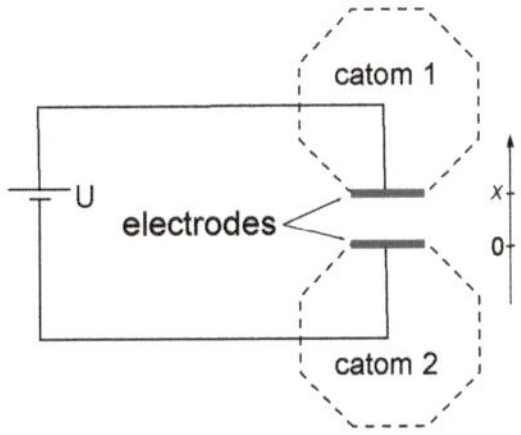

Figure 3. Principle of electrostatic actuator based on parallel capacitors.

Having the relationship that links the applied voltage U and the electrostatic force F as described in Equation (2), it is easy to derive the required condition to latch two catoms to each other. For this, let us consider the four configurations presented in Figure 4. They represent the possible positions between two catoms in latching condition and based on a FCC latice. Among these positions, two are of interest: Case A and Case B. Case A is the critical case where the entire weight of one catom has to be overcame by the connection, whilst case B represents the worst case where the moment to be overcame is maximal. Cases C and D represent less force and less moment than the two first cases, therefore they will not be considered here.

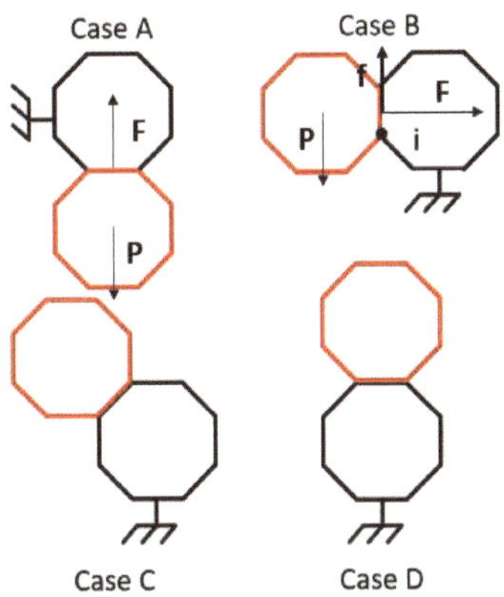

Figure 4. Different possible configurations of catoms based on face-centered cubic (FCC) lattice.

First let us consider case A. The minimum force to maintain the catom is straightforward from the Newton's first law.

$$F = mg \tag{3}$$

where m is the weight of the catom and g is the gravity. By combining Equations (2) and (3), considering that there is an air gap between the two electrodes, the condition in term of voltage that guarantees the latching is obtained:

$$U \geq \sqrt{\frac{2mgx_{air}^2}{\varepsilon_0 \varepsilon_r A}}. \tag{4}$$

Let us now consider case B. Still using Equations (2) and (3), the required moment around point I (Case B in Figure 4) to maintain the catom latched is derived:

$$\frac{Fc}{2} \geq mgr, \tag{5}$$

where c is the connector side and r the external radius of the quasi-spherical shape that forms the catom. Combining Equations (2) and (3) leads to the second condition in term of voltage required to make sure that the catom will not roll down when latched:

$$U \geq \sqrt{\frac{4rmgx_{air}^2}{c\varepsilon_0\varepsilon_r A}}. \tag{6}$$

From the two conditions in Equations (4) and (6), the final condition to be ensured in order to guarantee both case A and case B, and consequently case C and case D, is:

$$U = \max\left\{\sqrt{\frac{2mgx_{air}^2}{\varepsilon_0\varepsilon_r A}}, \sqrt{\frac{4rmgx_{air}^2}{c\varepsilon_0\varepsilon_r A}}.\right\} \tag{7}$$

2.3. Simulations

To perform the simulations, numerical values for the further realization are used. The material to be used is the VISIJET M3 Crystal material, a plastic used for 3D printing. From this material and using Autodesk Inventor 2019 CAD software, we estimate the weight of the catom to be 500 mg. From this weight and using Equation (3), the force has to be greater than F = 5.10 mN in order to lift a catom in vertical position (case A). To derive the necessary conditions in term of applied voltage and distance between electrodes that can provide this amount of force, we start by simulating the general governing equation of electrostatic actuation as described in Equation (1). To this aim, some hypothesis are made: (i) ε_r is taken equal to 1 as we assume that only air insulates the two electrodes, and (ii) the electrodes are perfectly planar. The resulting force versus the applied voltage and the gap between electrodes is given in Figure 5.

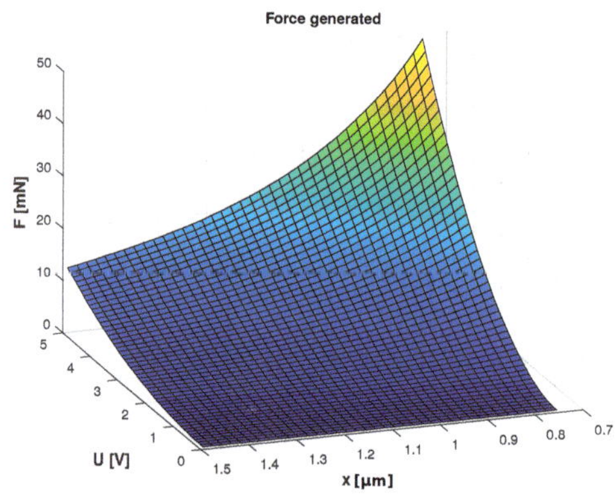

Figure 5. Force generated versus voltage and distance between two electrodes.

As expected from Equation (2), Figure 5 clearly reveals that the distance between the electrodes has an important role regarding the generated force: the further distance between two electrodes is, the less generated force will be. In other words, few hundreds of nanometers of additional distance will drastically decrease the generated force. In counterpart, reducing the electrodes gap in order to increase the amount of force for a given voltage has a limitation. Indeed, one has to take into account the dielectric strength (breakdown voltage) of the material isolating the electrodes. When subjected to a voltage higher than its dielectric strength, the material breaks and electrical arcs will appear between the two electrodes and will reduce the generated force to zero. Furthermore when the material is broken, the electrodes can not be used anymore. As an example, let us consider a gap between x_{air} two electrodes equals to 10 µm. Doing so, it is easy to determine the required force to lift the catom respecting the two previous conditions. Considering as example a voltage of 65.8 V, one can obtain from the intersection of the two curves illustrated in Figure 6: the force generated by the weight of the catom and the electrostatic force attracting each other the electrodes separated by a distance of 10 μm. For example, the air has a dielectric strength of 3.0 Mv/m. Hence, a voltage of 30 V can cause breakage of the air between the two electrodes distanced of 10 μm. Thus, another insulation material has to be used or added to separate the electrodes in the suggested electrostatic actuators. In this regard, SiO_2 (silica) material is of particular interest since it has a dielectric strength of 40 MV/m that requires 390 V of voltage to reach the breakage [21]. In addition, the integration and deposition of this material to the electrodes is standard using clean room facilities and thus a good flatness of the final electrodes will be ensured. With this SiO_2 layer on each electrode, the final gap between two electrodes is: $x = x_{air} + x_{ins}$.

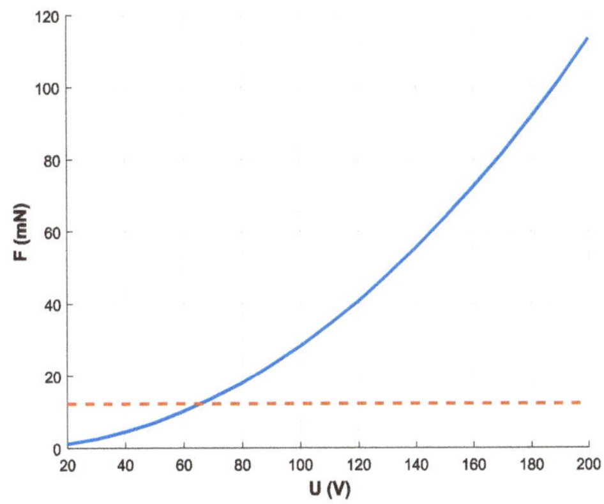

Figure 6. Force generated versus voltage considering a gap of x = 10 μm.

3. Fabrication of the Catoms

As from the previous section, ensuring a fully planar gap between two electrodes is a key challenge for electrostatic actuation. To ensure a high quality planarity, we combined micro-fabrication technique and a high-resolution 3D printing. Micro-fabrication technique was used to fabricate electrodes in order to guarantee a high level of flatness whilst 3D printing was used to print a catom where grooves were expected to host the electrodes.

3.1. Realization of the Catom Structure

As quasi-spherical shape is not easy to fabricate with conventional process, a 3D printer is used for fast and cheap prototypes it. It consists of a PolyJet SD 3500 printer (3D SYSTEMS company, South Carolina, USA). The printer uses a Visijet M3 Crystal material to print the design itself while wax is used as a support material. The model is constructed by adding material layer by layer. After printing, the model is placed inside an oven to remove the wax material. To reduce the catom weight as much as possible, and to make easy the further electrodes wiring, holes are created within the catom. The resulting structure after printing is presented in Figure 7 (left). The printed catom presents a weight of m = 373 mg.

Figure 7. Printed catom's structure and catom with glued electrode.

3.2. Clean Room Process

As pointed in Section 2.3, the distance between electrodes has very high impact on the force generated by the catoms. Even a small difference in the gap can drastically reduce the produced force. To overcome this limitation, a high quality of flatness of the electrodes is required. To reach this, clean room technologies are the suitable solution since they can guarantee the required flatness of the surface. It is worth to notice that micro-fabrication process guarantees a very high level of flatness since the deposition process is controlled with a maximum error of 3% over the whole wafer. Reported to the size of one electrode this error becomes smaller at around 0.2%. Based on these technologies, the flow chart process given in Figure 8 is developed in this work to fabricate the electrodes and the deposition of the additional insulator based on SiO_2.

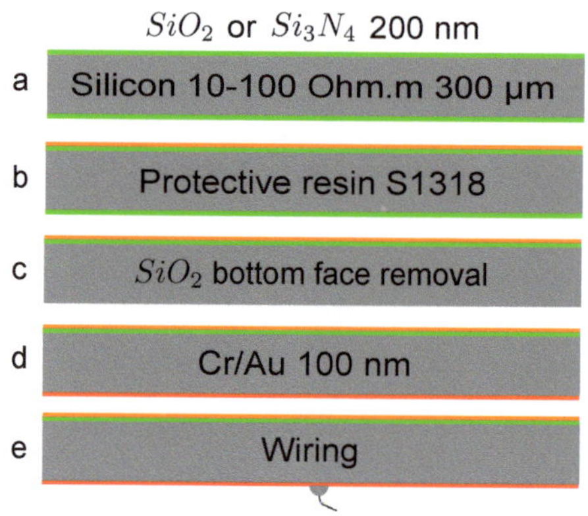

Figure 8. Flow chart of the clean room process used to fabricate the electrodes.

The process starts by growing an insulation layer of 200 nm of SiO_2 on both sides of a low resistant wafer with a thickness of 300 μm using a thermal oxidation oven (see Figure 8a). The oxidized layer, also referred to as insulator or insulation layer, has a 1.5% error margin on its thickness. Then, the SiO_2 on the bottom face is removed. To do so, the top face is protected with a S1318 resin (see Figure 8b) and the wafer soaks in a BHF bath (see Figure 8c). Then, the bottom side of the wafer is metallized using a cathode-ray pulverization machine to apply first a 20 nm layer of chrome and then a 80 nm layer of gold (see Figure 8d). The metal layer has two main functions: (i) first, it is used to act as soft layer to glue the wire without (or with very few) oxydation, (ii) second, it permits to have a better distribution of the electric field in the silicon layer allowing a more homogeneous distribution of the charge. In the last step, the wafer is cut following a square pattern of 8 mm side, which represents the size of the electrodes. The electrodes are chosen in the middle of the wafer to further minimize the flatness error. Then a silver paste is applied to connect an insulated wire of a diameter of 0.2 mm on the metallized side (see Figure 8e). This wire is chosen to be very flexible and light in order to neglect its effects during the experiments. Finally, electrodes and their wires are glued on the catom structure as shown in Figure 7 (right). SiO_2 material is privileged as insulator because of its high dielectric strength and its high dielectric constants. Si_3N_4 can also be used but the quality of its deposition onto the catom surface is challenging.

To do the assembly and fabricate the final catom, a small amount of superglue is spread out along the catom's grooves, then the electrodes are glued. This operation is done manually since the size of the electrodes (side of 8 mm) and the structure (diameter of 2 cm) are sufficiently large to be manipulated by manual tweezers. In this case, there is no need to: (i) master the glue, (ii) master applied force or (iii) use an automated assembly technique.

4. Experimental Validation

4.1. Presentation of the Experimental Benchmark

The experimental benchmark used to validate the fabricated catoms is depicted in Figure 9. This setup is used to characterize the generated force when two catoms latch each other. The experimental setup is composed of:

- a tensile test machine equipped with a load cell having a force capability up to 2 N with a resolution of 3 mN. The top part of the machine is the moving part which produces a tensile force. The bottom part is fixed and connected to a load cell which acts as the sensor.
- two catoms fixed on the jaws of the tensile machine. One catom is fixed on the top of the machine while the second is fixed on the bottom of the machine.
- a generator that can provide a voltage up to 200 V allowing to supply the electrodes of the catoms.
- and a computer and acquisition system used to acquire signals from the tensile machine, to visualize and to record data, and to analyze them.

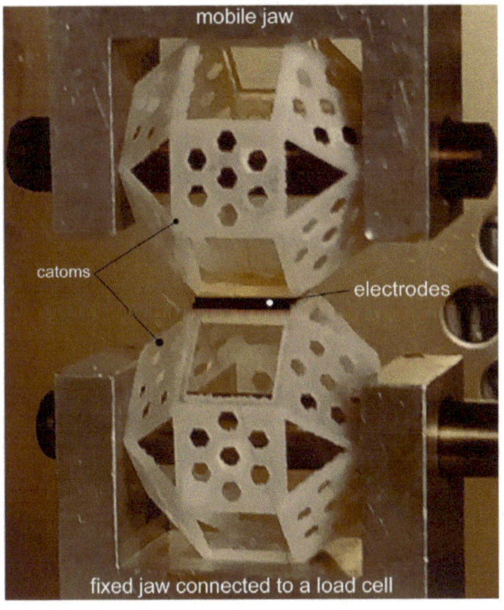

Figure 9. Experimental bench.

The experimental setup are placed on an anti-vibration table to limit the influence of mechanical vibration on the load cell of the tensile machine.

4.2. Tests and Results

The latching capability of the catom is tested for different applied voltages. The idea consists to do a characterization of the interaction force between two catoms according to the applied voltage.

In other words, it aims to derive a relationship between the applied voltage and generated force. For each test, the same experimental protocol given below is followed:

1- The first catom is clamped to the bottom jaw of the tensile machine.
2- The second catom is placed on top of the first one.
3- The two catoms are powered to latch them together.
4- The second catom is clamped to the top jaw of the tensile machine while latching to the first catom.
5- The tensile test begins by moving up the top jaw of the tensile machine with a constant speed of 1 μm/s.

The experiments were carried out using several sets of catoms and at different voltages. Since the test is not destructive, we were able to take several measurements for each voltage values and each catom sets. Figure 10 shows a typical interaction force recorded by the load cell of the tensile machine. Two distinct phases can be observed. At the beginning, the force grows quickly until the breakout force is reached. Then instead of dropping to zero instantly, the force decreases slowly. This is the effect of electrostatic force which decreases when the distance between two electrodes increases.

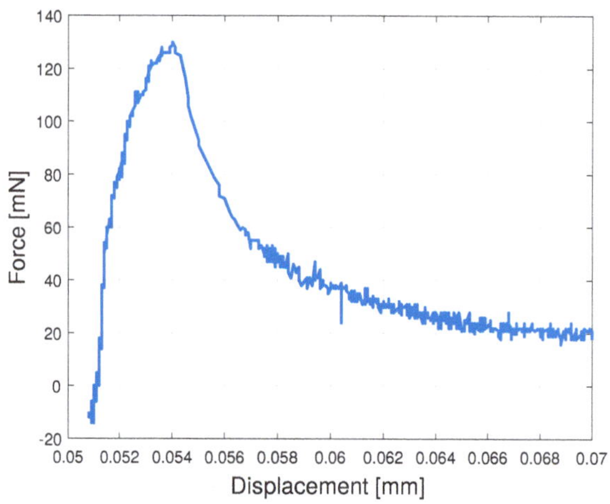

Figure 10. Breakout Force measured during the tensile test at 170 V.

Here, only the peak is of interest because we are only studying the latching mechanism. The measurement is repeated 150 times for different voltages between 100 V and 200 V. The minimum voltage is 100 V because below this point, the measurement noise becomes important compared to the produced force of the mechanism. Each test gave off a force value depending on the applied voltage and the results are compiled in Figure 11. This curve particularly shows the standard deviation of the measured force per voltage and the least square identification used to verify the model given in Equation (2). The red curve is the same as the curve given in Figure 6 where the air gap value is adapted to fit the model. These results are discussed in the next subsection.

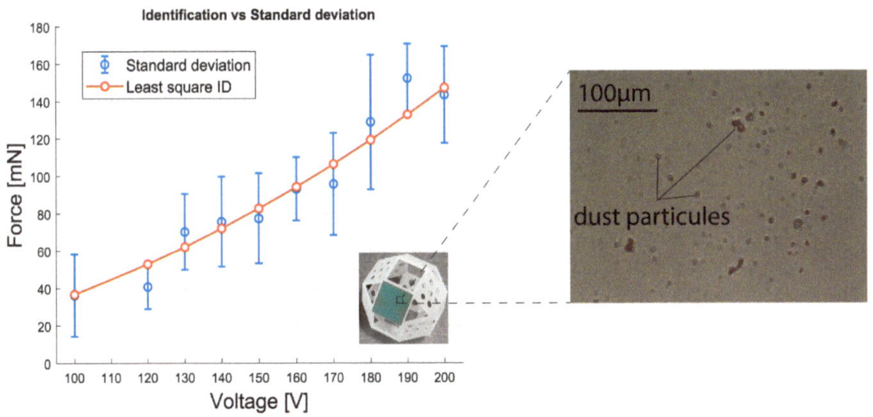

Figure 11. Standard deviation curve versus least square identification method.

4.3. Discussions

When comparing the experimental values with the theory for a gap of 600 nm (equals to the sum of the SiO_2 thicknesses of the two antagonist electrodes), a huge difference is observed. The theory exhibits 60 mN of force for 100 V of voltage while the experiment gives a generated force of approximately 40 mN. To analyze the difference, let us go back to Equation (2). Thanks to the clean room process of fabrication, parameters x_{ins} and A have very negligible errors compared to the theory. Hence the remaining parameter is x_{air} which is the air gap. This air gap can come from two different factors: (i) a tilt or parallelism error between the electrodes, (ii) dust particles. Regarding the slight tilt, it come from the gluing or the position of a catom according to another one, but the generated force can compensate it since the electrodes are planar and one catom is always free. However dust that comes from the environment cannot be compensated because it introduces a physical gap that increases the distance between two electrodes and as a consequence decreases the generated force. To support this hypothesis a zoom of an electrode is shown in Figure 11 where dust particles can be seen. Assuming such dust, using a parametric identification technique, we found that an air gap of about 12 µm potentially exists between the two electrodes.

Because dust size is approximately 11 µm, this assumption well fits to the experimental results in Figure 11. Regarding the high standard deviation in this curve however, its cause is due to the difficulty to set the same initial condition each time we start a new test. When the two catoms split, they slightly move due to the used 'pin-jaws'. In addition, the pin-jaws have a certain flexibility which allows to easily re-latch the catoms between each test, and thus implies that the catoms are not in the exact same position than before. Moreover, in the meantime, it is also possible that additional dust come to or disappear from between the two electrodes and consequently change the air gap previously mentioned. Finally the electrodes are also fragile: when subjected to shocks they can stop working due to the dielectric layer breaking and the prolonged usage with repeated latching and unlatching can wear off the electrodes.

In spite of these difficulties, the identified relation between the applied voltage and the generated force is important since it allows to predict the sufficient amount of force to achieve a given configuration. Keeping in mind this relation, we performed different configurations. The results shown in Figure 12 are very promising in terms of latching capabilities of a catom. Indeed Figure 12a,b show that one can achieve the conditions given in Section 2.2. On the other hand Figure 12c,d reveal that when a maximum voltage of 200 V is applied, the catom's latching capabilities are strengthen. Such high voltage contributes for the stability of the latching for overhanging situation or when a heavy object (75 times heavier than one catom) have to be supported by the catom.

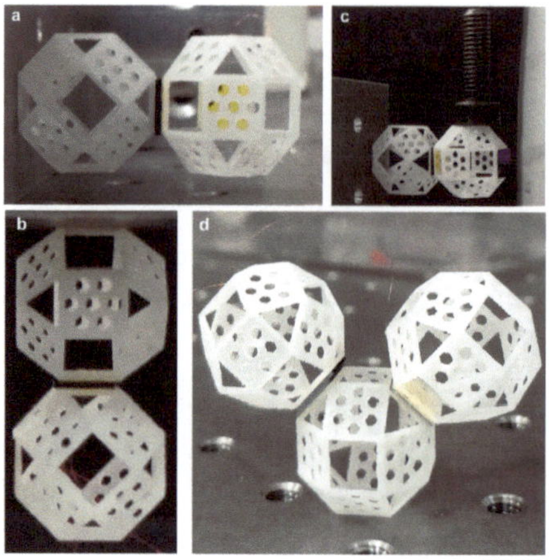

Figure 12. Experimental validation of different catom's configurations. (**a**) Catom latching in overhang configuration (Case B). (**b**) Catom latching in vertical configuration (Case A). (**c**) Catom latching in overhang configuration while lifting a heavy weight. (**d**) Two catoms latching following Case C.

5. Conclusions

This paper presented the design of a new module for programmable matter. Called catom, the module was typified by a quasi-spherical shape with face-centered cubic lattice and had centimeter scale. Electrostatic actuation was studied for the latching capability of the catom. After describing the catom's geometry, we provided the latching conditions considering low energy, miniaturization and sufficient force constrains. The fabrication of the catoms was afterwards presented and simulation and experiments were carried out. The results demonstrated that the latching was strong enough under various conditions.

Future works consist in improving the robustness of the electrodes regarding their wearing off by adding a protecting layer and/or utilizing a metallic wafer. Future works also include the study of flexible electrodes for a better fitting with the quasi-spherical shape. On the other hand, the used electrostatic actuation will also be studied and extended such that one catom can move relative to another catom instead of only working in latching mode. In this future work, additional parameters such as the friction and the catom inertia will have to be considered in order to find the conditions of rolling/toplling of a catom relative to another one. Finally, perspective works will also include the integration of a super capacitor or a non-contact source to supply each catom.

Author Contributions: Authors contributions: R.C. did the investigation and A.M.-O., M.R. and P.L. are the supervisers of the works. All authors have read and agreed to the published version of the manuscript.

Funding: This work is supported by the French "Investissements d'Avenir" program, ISITE-BFC project (ANR-15-IDEX-03) and the EIPHI Graduate School (contract "ANR-17-EURE-0002"). This work is also partly supported by the french RENATECH network and its FEMTO-ST technological facility.

Acknowledgments: The authors would like to express their thanks to the french "Investissements d'Avenir" program and the EIPHI-UBFC Graduate school program. We would like also to thank Dr. Xavier GABRION for helping us with the experimental characterization.

Conflicts of Interest: The authors declare no conflict of interest.

References

1. Toffoli, T.; Margolus, N. Programmable matter: Concepts and realization. *Phys. D* **1991**, *47*, 263–272. [CrossRef]
2. Yim, M.; White, P.; Park, M.; Sastra, J. Modular Self-Reconfigurable Robots. In *Encyclopedia of Complexity and Systems Science*; Springer: New York, NY, USA, 2009; pp. 5618–5631.

3. Fukuda, T.; Ueyama, T.; Kawauchi, Y.; Arai, F. Concept of cellular robotic system (CEBOT) and basic strategies for its realization. *Comput. Electr. Eng.* **1992**, *18*, 11–39. [CrossRef]
4. Satoshi, M.; Eiichi, Y.; Kohji, T.; Kurokawa, H.; Kammimura, A.; Kokaji, S. *Hardware Design of Modular Robotic System*; IEEE IROS: Piscataway, NJ, USA, 2000.
5. Romanishin, J.W.; Gilpin, K.; Rus, D. M-Blocks: Momentum-Driven Magnetic Modular Robots. In Proceedings of the IEEE/RSJ International Conference on Intelligent Robots and Systems, Tokyo, Japan, 3–7 November 2013; pp. 4288–4295.
6. Available online: http://www.cs.cmu.edu/~claytronics/hardware/helium.html (accessed on 3 June 2020).
7. Garcia, R.F.M.; Hiller, J.D.; Lipson, H. *A Vacuum-Based Bonding Mechanism for Modular Robotics*; IEEE ICRA Workshop: Piscataway, NJ, USA, 2010.
8. Kotay, K.; Rus, D.; Vona, M.; McGray, C. *The Self-Reconfiguring Robotic Molecule*; IEEE: Piscataway, NJ, USA, 1998.
9. Murata, S.; Kurokawa, H.; Yoshida, E.; Tomita, K.; Kokaji, S. *A 3-D Self-Reconfigurable Structure*; IEEE ICRA: Leuven, Belgium, 1998; pp. 432–439.
10. Jorgensen, M.W.; Ostergaard, E.H.; Lund, H.H. *Modular ATRON: Modules for a Self-Reconfigurable Robot*; IEEE IROS: Piscataway, NJ, USA, 2004; Volume 2, pp. 2068–2073.
11. Chirikjian, G. *Kinematics of a Metamorphic Robotic System*; IEEE ICRA: Leuven, Belgium, 1994; pp. 449–455.
12. Will, P.; Castano, A.; Shen, W. Robot modularity for self-reconfiguration. *Proc. SPIE* **1999**, *3839*, 236–245. [CrossRef]
13. Belisle, R.; Yu, C.; Nagpal, R. Mechanical design and locomotion of modular-expanding robots. In Proceedings of the IEEE 2010 International Conference on Roboticsand Automation Workshop, Modular Robotics: State of the Art, New York, NY, USA, 3 May 2010; pp. 17–23.
14. Karagozler, M.E. Design, Fabrication and Characterization of an Autonomous, Sub-Millimeter Scale Modular Robot. Ph.D. Thesis, CMU, Pittsburgh, PA, USA, 2012.
15. Gilpin, K.; Knaian, A.; Rus, D. *Robot Pebbles: One Centimeter Modules for Programmable Matter through Selfdisassembly*; IEEE ICRA: Leuven, Belgium, 2010; pp. 2485–2492.
16. Piranda, B.; Laurent, G.; Bourgeois, J.; Clévy, C.; Mobes, S.; Le Fort-Piat, N. A new concept of planar self-reconfigurable modular robot for conveying microparts. *Mechatronics* **2016**, *23*, 906–915. [CrossRef]
17. Ostergaard, E.H.; Kassow, K.; Beck, R.; Lund, H.H. Design of the ATRON lattice-based self-reconfigurable robot. *Auton. Robot.* **2006**, *21*, 165–183. [CrossRef]
18. Pengwang, T.E.; Rabenorosoa, K.; Rakotondrabe, M.; Andreff, N. Scanning micromirror platform based on MEMStechnology for medical applications. *Micromachines* **2016**, *7*, 24. [CrossRef] [PubMed]
19. Available online: https://projects.femto-st.fr/programmable-matter/teams (accessed on 3 June 2020).
20. Piranda, B.; Bourgeois, J. Designing a quasi-spherical module for a huge modular robot to create programmable matter. *Auton. Robot.* **2018**, *42*, 1619–1633. [CrossRef]
21. Available online: http://eesemi.com/sio2si3n4.htm (accessed on 3 June 2020).

actuators

MDPI

Article

Development of Haptic Stylus for Manipulating Virtual Objects in Mobile Devices

Dong-Soo Choi, In-Ho Yun, Tae-Hoon Kim, SangKyu Byeon and Sang-Youn Kim *

Interaction Laboratory, Advanced Research Technology Center, Interdisciplinary Program in Creative Engineering, Korea University of Technology and Education, 1600, Chungjeol-ro, Dongnam-gu, Cheonan 31253, Korea; mycds88@koreatech.ac.kr (D.-S.C.); dogooly@koreatech.ac.kr (I.-H.Y.); rlaxogns94@koreatech.ac.kr (T.-H.K.); skbyeon89@gmail.com (S.B.)

* Correspondence: sykim@koreatech.ac.kr; Tel.: +82-41-560-1461

Received: 24 March 2020; Accepted: 12 April 2020; Published: 14 April 2020

Abstract: In mobile devices, the screen size limits conveyance of immersive experiences; haptic feedback coupled with visual feedback is expected to have a better effect to maximize the level of immersion. Therefore, this paper presents a miniature tunable haptic stylus based on magnetorheological (MR) fluids to provide kinesthetic information to users. The designed stylus has a force generation, force transmission, and housing part; moreover, in the stylus, all three operating modes of MR fluids contribute to the haptic actuation to produce a wide range of resistive force generated by MR fluids in a limited size, thereby providing a variety of pressing sensations to users. A universal testing machine was constructed to evaluate haptic performance of the proposed haptic stylus, whose resistive force was measured with the constructed setup as a function of pressed depth and input current, and by varying the pressed depth and pressing speed. Under maximum input voltage, the stylus generates a wide range of resistive force from 2.33 N to 27.47 N, whereas under maximum pressed depth it varied from 1.08 N to 27.47 N with a corresponding change in voltage input from 0 V to 3.3 V. Therefore, the proposed haptic stylus can create varied haptic sensations.

Keywords: miniature stiffness display; MR fluids; haptic; multiple mode

1. Introduction

Because of advancements in mobile hardware technology, interaction with mobile virtual objects has become increasingly popular with most users owning a smart phone. Mobile virtual reality includes a variety of graphical content from a simple static object to one having complex dynamic behavior. A user who manipulates and interacts with these objects wants to haptically "feel" not only the presence of objects but also their material properties. As the demand for realistic interaction with virtual objects increases, the importance of haptic technology also increases to ensure that the ultimate level of immersion is conveyed to users.

Haptic sensation consists of kinesthetic and tactile feelings. A kinesthetic feeling refers to sensory data obtained through receptors of joints, muscles, ligaments, etc. The tactile feeling is a cutaneous sensation obtained from receptors of the skin. A user recognizes the stiffness of an object through kinesthetic information and discerns the texture of an object through tactile information. Thus, to convey a more realistic haptic sensation to users in a virtual environment, both tactile and kinesthetic information should be simultaneously presented to users.

Recently, many mobile devices have used styluses to help users tap on the screen and allow them to delicately manipulate virtual contents. In virtual reality, while visual information is the most dominant sensory input for perceiving virtual objects, haptic information coupled with visual information increases the sense of reality. Therefore, many researchers have focused on tactile modules that stimulate the skin of users [1–7]. Lee et al., developed a haptic pen using a touch sensor and solenoid coil to create a sense of contact with virtual objects [1]. Kyung and Park presented a pen-like haptic interface, which consists of a compact pin-array tactile module and a vibrating module (Ubi-Pen), and verified its haptic performance [2,3]. Liao et al., developed a pen-type haptic interface device and evaluated its usefulness [4]. Cho et al., presented a haptic stylus to generate auditory-tactile feedback that provides realism in handwriting or drawing tasks [5]. Wang et al., suggested an electro-vibration based haptic pen which can create continuous tactile feedback [6]. Arasan et al., developed a vibrotactile pen for generating two haptic effects (static and dynamic vibration) and conveying them to users [7]. Although these pens can haptically simulate the roughness of a target object and can emulate button sensation using vibration, it is not easy to represent the stiffness of the virtual object through tactile feedback.

Many researchers have shown interest in incorporating kinesthetic effects in haptic styluses. Kamuro et al., suggested a pen-type hand held haptic interface conveying kinesthetic feeling to users [8]. Tian et al., proposed a haptic stylus which creates kinesthetic and tactile sensations [9]. Kianzad and MacLean developed a magic pen that generates kinesthetic sensation using a ball-point drive mechanism [10]. Kara and Patoglu suggested a haptic stylus having variable tip compliance [11]. Even though these haptic styluses directly provide kinesthetic information (for example, force or distributed pressure), actuation mechanisms are too bulky and heavy to be inserted into tiny styluses.

Therefore, the magnetorheological fluid (MR fluid), whose mechanical properties can be changed according to an external magnetic field, has been studied as a material for pen-type haptic devices to convey kinesthetic sensation to users [12,13]. MR fluids based haptic pen was developed to interact with an image on a mobile terminal [12]. Another haptic pen for creating kinesthetic sensation using a hybrid actuator (MR actuator with a voice coil motor) was introduced [13]. They used two modes (a flow mode and a shear mode) of the available three operating modes of MR fluids to increase haptic resistive force. Although their research works greatly help users to understand the material property of virtual objects by using kinesthetic information, the range of haptic force can be further expanded if the squeeze mode is added to the haptic pen. The reason being that squeeze mode has the greatest effect on the generation of resistive force among these three operating modes in MR fluids [14–18]. In this paper, we propose a new design of tiny haptic stylus, in which all three modes (a shear mode, flow mode, and squeeze mode) of MR fluids can contribute to the haptic actuation, to increase the magnitude of haptic force to the level where users are truly "satisfied" in a small size. The proposed tiny haptic stylus is expected to be useful for interaction with virtual objects because of its simple structure and haptic performance.

2. Design of A New Kinesthetic Stylus

2.1. Overall Structure

Figure 1 shows the structure of a proposed haptic stylus which consists of three parts: (1) force generation, (2) force transmission, and (3) a housing part. The force generation part, which plays a role in creating resistive force, includes a T-shape rod, a spring, an O-ring, a bottom plate, a spacer, a plunger, a solenoid coil, a magnetic shelter, and a top plate. The force transmission part is composed of a pen tip and hollow shaft. The spring is used to restore the compressed force generation part to its initial position and the T-shaped rod transmits the resistance force from the force generation part to the force transmission part. One end of the T-shaped rod is connected to the hollow shaft, and the other end of the T-shaped rod is inserted into the hole of the bottom plate with a spring. The hollow shaft acts as the central axis of the vertical movement of the proposed module. The housing part consists of an upper and a lower housing. The upper housing is designed to minimize magnetic flux leakage when voltage input is applied to the solenoid. The force generation part is inserted into the upper housing. The T-shaped rod is connected to one end of the hollow shaft, and the pen tip is inserted into the other end of the hollow shaft. The lower housing covers the force transmission part and is then inserted into the upper housing.

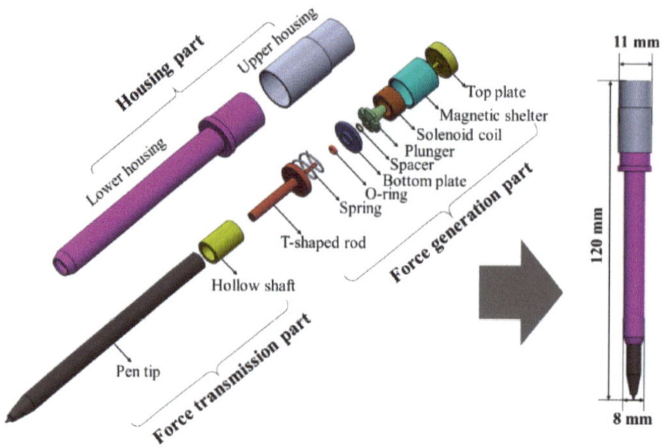

Figure 1. Schematic of the proposed kinesthetic stylus.

2.2. *Operating Principle*

The key components of the proposed haptic stylus are MR fluid (MRF) and the solenoid. The MR fluid is filled in the magnetic shelter that includes the solenoid. The solenoid coil creates magnetic fields to control the stiffness of MR fluids. The magnetic shelter provides a path along which the magnetic field induced by the solenoid coil flows. Figure 2 shows the operating principle of the proposed haptic stylus. When there is no voltage in the solenoid, the MRF is in the fluid state. Therefore, when a user holds the proposed haptic stylus and pushes the screen of a mobile device with the pen tip, the haptic stylus is softly pressed, and the user can feel the softness. If an input voltage is applied to the solenoid coil in the proposed module, the magnetic particles in the MRF form chains and the MRF becomes semisolid. During this change, since the solidification extent is proportional to the applied magnetic field, the force generation part in the haptic stylus can create various haptic sensations. Therefore, when a user interacts with mobile virtual contents using the haptic stylus, he/she can sense a variety of haptic sensations.

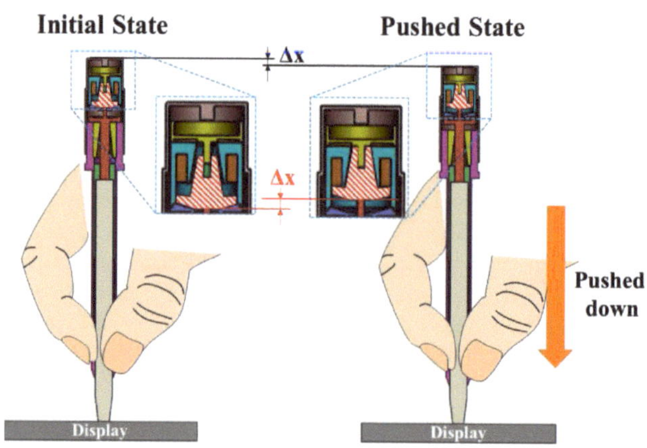

Figure 2. Operating principle of the proposed haptic stylus.

2.3. *Optimization of the Haptic Stylus*

For maximizing the haptic performance of the proposed stylus, we need to not only optimize the solenoid and the gap distance between magnetic shelter and the plunger but also reduce the friction between the plunger and the top plate. First, we optimize the diameter of the solenoid coil to maximize resistive force with less power consumption. We simulated the solenoid coil assuming that this module consumes a power of 0.6 W under an input voltage of 3.3 V. The resistive force of the proposed haptic stylus is highly related to the magneto-motive force of the solenoid coil. The magneto-motive force (F) of the solenoid coil and the power consumption are expressed by Equation (1) [19]:

$$F = \frac{NVA}{\rho\, l_t} \quad (1)$$

In Equation (1), N is the number of turns of the solenoid coil, V is the input voltage, A is the area of pure conductor, ρ is the resistivity of copper (1.72×10^{-8} Ωm), and l_t is the total length of the solenoid coil. Because we used a multilayer solenoid coil in the proposed haptic stylus, the area (A) of pure conductor and total length (l_t) of the copper wire are expressed by Equations (2) and (3), respectively.

$$A = \frac{\pi (d_w)^2}{4} \quad (2)$$

$$l_t = 2\pi r_m N \quad (3)$$

In Equations (2) and (3), π is the ratio of a circle's circumference to its diameter, d_w is a diameter of the wire of the solenoid coil, and r_m is the average radius of the solenoid coil.

The power consumption (P) of the solenoid coil is calculated by the following Equation:

$$P = \frac{V^2 A}{\rho l_t} \quad (4)$$

Figure 3 shows the simulation result of the magneto-motive force (MMF) of the solenoid coil according to its wire diameter. The black line is the computed magneto-motive force by varying the wire diameter, and the red line is the magneto-motive force with a fixed power consumption of 0.6 W. From the result, the wire diameter of the solenoid coil was selected as 0.11 mm.

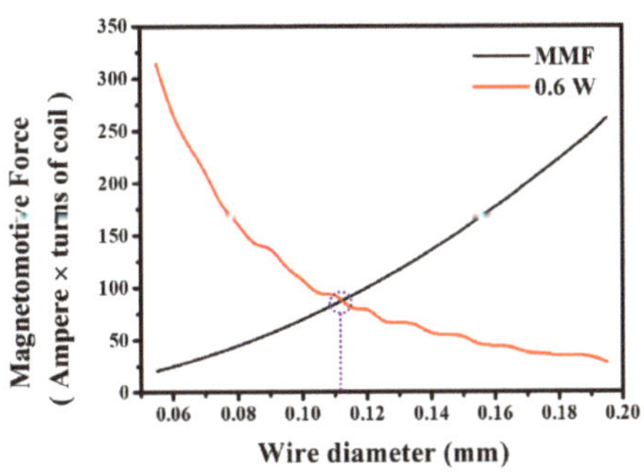

Figure 3. Simulation result of the solenoid coil: magneto-motive force vs. wire diameter.

As mentioned earlier, there are three operating modes in MRF: (1) a flow mode (MR fluids flow due to a pressure gradient between the two stationary plates), (2) a shear mode (MR fluids flow between two plates that are moving relative to one another), and (3) a squeeze mode (MR fluids flow between the two plates that are moving in a direction that is perpendicular to their planes) [14–18]. To maximize the resistive force in a small device, the proposed haptic stylus should be fabricated in a structure in which all three operation modes of MR fluids can contribute to the actuation. To do this, an FEM simulation was conducted using a commercialized software (Maxwell 2D v14, Ansoft, Pennsylvania, USA). Figure 4 shows the cross-sectional view of the assembled force generation part and its flux path simulation result. The solenoid coil is attached and fixed to the inside of the magnetic shelter. When the T-shaped rod is pressed, this pressing force causes the plunger to move upward. At this time, the magnitude of the resistive force can be controlled by changing the state of MR fluid. From the results (Figure 4), we found that the magnetic flux flows well without magnetic saturation or leakage in the proposed design. As the plunger moves down, the distance between itself and the declined plane of the magnetic shelter becomes closer and the MR fluid is pressed (squeeze mode). The squeezed MR fluid flows through the gap between the plunger and the magnetic shelter (flow mode). Meanwhile, the movement of the plunger causes relative motion between the two slopes (the plunger and the magnetic shelter) and this relative motion creates shear stress (shear mode). Thus, we found that the proposed haptic stylus is designed to use all three operating modes of the MR fluids so that it can generate sufficient resistive force in a small size.

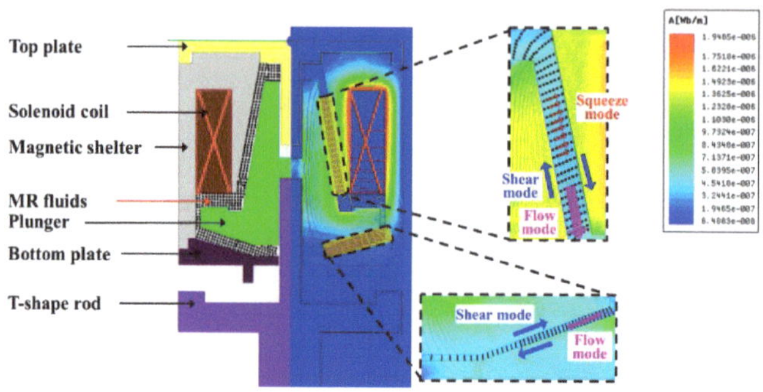

Figure 4. Cross-sectional view of the force generating part and its simulation result.

The gap distance between the magnetic shelter and the plunger is another important factor to maximize the haptic performance because the gap is filled with MR fluid. Thus, we simulated the resistive force (haptic performance) by varying the distance (y) between the magnetic shelter and the plunger (Figure 5). The distance (y) was varied from 1.3 mm to 1.8 mm at an interval of 0.1 mm. As shown in Figure 5, the result indicates that the resistive force decreases as the distance (y) increases. Therefore, it is better to decrease the distance between the magnetic shelter and the plunger. However, in the case of a distance of 1.3 mm, the normal distance between the magnetic shelter and the plunger when the haptic stylus is fully pressed (d = 1 mm) is too close ($(y - d) \cdot \sin\theta = (1.3 - 1) \cdot \sin(10°) = 0.05$ mm) considering the manufacturing error. Therefore, in this study, we selected the distance as 1.4 mm.

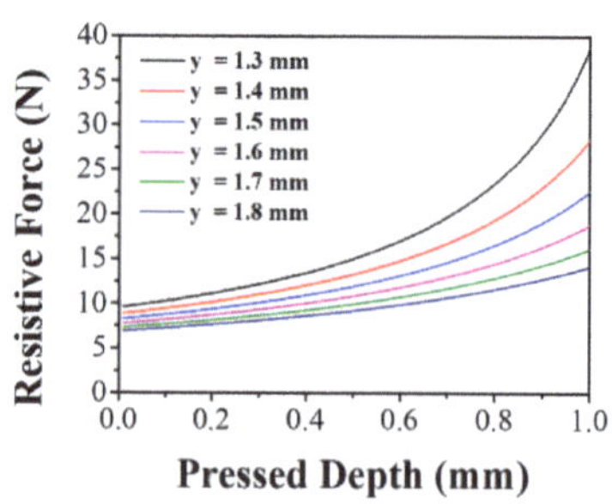

Figure 5. Simulation result of the resistive force of the haptic stylus by varying the distance between the magnetic shelter and the plunger.

To reduce the initial resistive force (when there is no current), we simplified the top plate as shown in Figure 6. Due to this simple design of the top plate, we could increase the area where the magnetic flux flows, resulting in the improvement of the haptic resistive force.

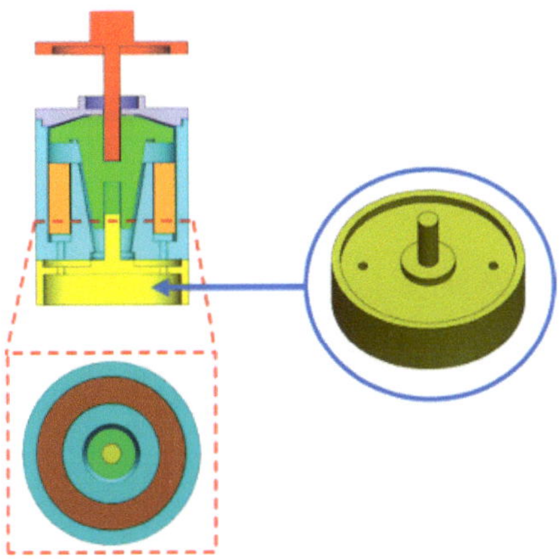

Figure 6. The simple design of the top plate.

2.4. Fabrication of the Tiny Haptic Stylus

Figure 7 shows the constructed haptic stylus consisting of the lower housing, upper housing, hollow shaft, T-shaped rod, spring, bottom plate, O-ring, spacer, plunger, solenoid coil, magnetic shelter, top plate, and the pen tip. The T-shaped rod is linked to one end of the hollow shaft, and the pen tip is connected to the other. The spring is located between one end of the T-shaped rod and bottom plate in order to create an elastic returning force. The end of the T-shaped rod that penetrates the bottom plate fits into the plunger's hole and is fully attached to the plunger. The bottom plate is connected to the magnetic shelter and MR fluid is filled into the magnetic shelter, and the plunger is then placed into the magnetic shelter. The commercial mini O-ring (T-SK818, 1.0 mm (diameter) × 0.4 mm (thickness)) was used with small friction and small size to prevent MR fluid leakage. The haptic stylus is completed by putting the lower housing toward the pen tip and fitting the upper housing to the top plate side.

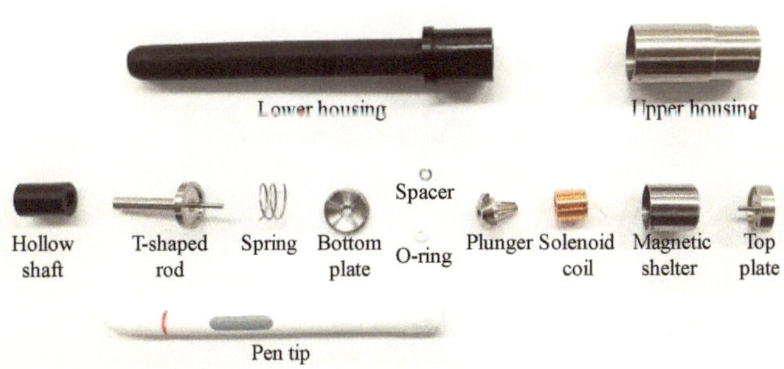

Figure 7. Components for the proposed tiny haptic stylus.

3. Experiment of the Tiny Haptic Stylus

3.1. Experimental Setup

Figure 8 shows an experimental setup to evaluate the haptic performance of the proposed tiny haptic stylus. The experimental setup is composed of a PC and a universal testing machine (UTM, Z0.5, ZwickRoell, Ulm, Germany) including a load cell (Xforce P 100 N, Zwick Roell, Ulm, Germany), and a power supply. In order to tightly fix the proposed haptic stylus, we fabricated a mold (a sample holder), whose shape is exactly matched with the upper housing of the haptic stylus. The sample holder was made of stainless steel (SUS 301) for minimizing the distortion or deformation. The haptic stylus was bonded with the sample holder, which was bolted on the UTM, by super glue. The measurement head was precisely moved in the vertical (z-axis) direction by the linear motor installed in the UTM. When the measurement head pushes down the haptic stylus, the resistive force is measured through the load cell. The measured resistive force and the position of the measurement head were delivered and were stored on the PC. The experiment was conducted under various constant speed (0.1 mm/s, 0.5 mm/s, 1.0 mm/s, 2.0 mm/s, 5.0 mm/s, 10.0 mm/s, and 15.0 mm/s) of the measurement head. We set the offset when the contact force (0.01 N) from the load cell was measured. The power supply was connected to the proposed haptic stylus to control its resistive force. We measured the resistive forces by changing the input current, measurement head's position, and varying the pressing speed.

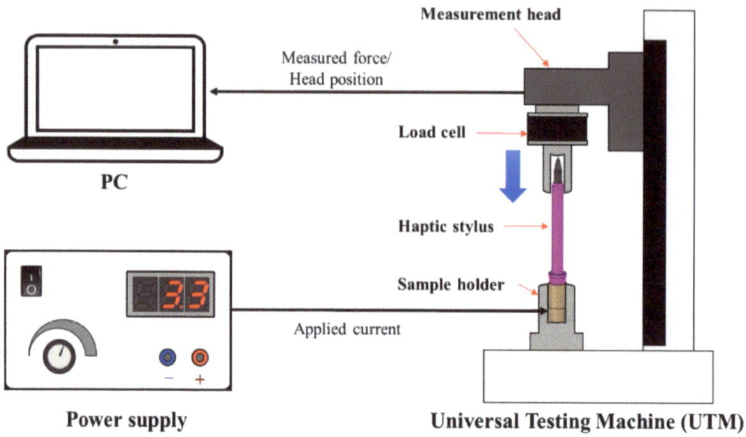

Figure 8. Experimental setup to evaluate the haptic performance of a proposed tiny haptic stylus.

3.2. Experimental Results

Figure 9a shows the experimental results of the measured force with varying indented depths (from 0.01 mm to 1.0 mm) and with fixing pressing speed (0.1 mm/s). In Figure 9a, the black line is the result when the current is removed. The initial resistive force of the proposed haptic stylus indicates approximately 1 N under the overall pressed depth because of the return spring and viscosity of MR fluid. The pink line is the result when the current (180 mA) is applied to the proposed tiny haptic stylus. Under the current input (180 mA), the measured resistive force gradually increases (from 2.33 N at 0.01 mm to 27.47 N at 1.0 mm) as the indented depth increases. For example, when the proposed haptic stylus is pressed 1.0 mm, the resistive force, which was initially 2.33 N, increases to 27.47 N in providing a current input of 180 mA. Theoretically, considering a human's Just Noticeable Difference (JND) of approximately 10% [20], the proposed haptic stylus allows users to distinguish more than 26 steps of resistive force. It means that the proposed haptic stylus can convey a variety of kinesthetic sensations when a user interacts with virtual reality contents in a mobile device. We also verified that the resistive force variation can be easily adjusted by changing the input current. Figure 9b shows the results of the measured resistive force with respect to the pressing depth (varied from 0 mm to 1 mm) and the pressing speed (0.1 mm/s, 0.5 mm/s, 1.0 mm/s, 2.0 mm/s, 5.0 mm/s, 10.0 mm/s, and 15.0 mm/s). The resistive force generated from the proposed haptic stylus was measured with a constant input voltage of 3.3 V (a consumed current is 180 mA). Even though the pressing speed increases, the measured resistive forces are almost similar to the reference data (0.1 mm/s).

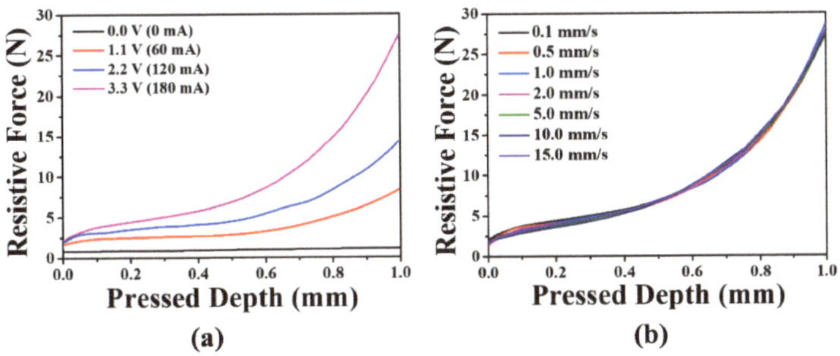

Figure 9. Measured resistive force by pressing depth with varying input current at a constant pressing speed of 0.1 mm/s (**a**) and with varying pressing speed at a constant input current of 180 mA (**b**).

We investigated the response time of the proposed haptic stylus by applying a step input to the haptic stylus and measuring the time taken to reach 90% of the reference signal. In Figure 10, the blue solid line is the measured resistive force when a step input is applied. The response time (67 ms) was calculated by subtracting t_1 (at the voltage is applied) from t_2 (at the measured resistive force reaches a point of 90 % of the 3.3 V reference). The results show that the response time of the proposed haptic stylus is enough to be used for conventional haptic applications [21–23].

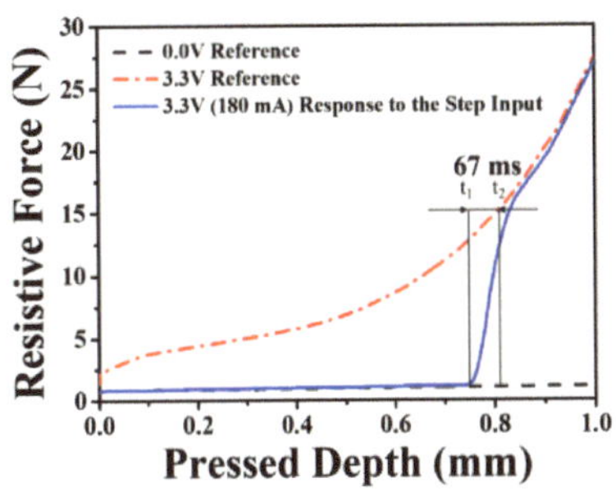

Figure 10. Measured response time of the proposed haptic stylus.

4. Conclusions

A tiny haptic stylus based on MR fluids was presented as an interaction device for manipulating virtual reality contents in mobile devices. Compared to conventional haptic styluses, the proposed device can produce a large kinesthetic force in a small size. To maximize the large kinesthetic force in a small-sized stylus, the proposed design includes all three operation modes of MR fluids. We verified that the resistive force's variation of the proposed haptic stylus was approximately 25.14 N (2.33 N–27.47 N) when the indented depth is varied from 0.01 mm to 1 mm. Moreover, we have found that the input voltage can change the resistive force of the proposed haptic stylus. Judging from the results, the proposed haptic stylus could be used as an interface device for manipulating virtual content in mobile devices. Some examples include touching rocks or cushion in virtual reality content; or conducting palpation training for medical students as medical education content. In order to maximize the performance of the proposed haptic stylus, we will develop the haptic rendering method and conduct a user study using the proposed haptic stylus. Furthermore, it may be possible to generate various haptic sensations by including a displacement or pressure sensor.

Author Contributions: Conceptualization, D.-S.C., S.B. and S.-Y.K.; Data curation, and S.B.; Formal analysis, D.-S.C. and T.-H.K.; Investigation, I.-H.Y. and S.B.; Visualization, D.-S.C.; Writing—original draft, S.-Y.K.; Writing—review and editing, S.-Y.K.; Authors 1 D.-S.C. and 2 I.-H.Y. contributed equally. All authors have read and agreed to the published version of the manuscript.

Funding: This work was supported by Priority Research Centers Program through the National Research Foundation of Korea (NRF) funded by the Ministry of Education, Science and Technology (NRF-2018R1A6A1A03025526). The study was also supported by the Technology Innovation Program (10077367, Development of a film-type transparent /stretchable 3D touch sensor haptic actuator combined module and advanced UI/UX) funded by the Ministry of Trade, Industry, and Energy (MOTIE, Sejong, Korea).

Acknowledgments: The authors thank the Cooperative Equipment Center at Koreatech for assistance with Universal Testing Machine (UTM) experiments.

Conflicts of Interest: The authors declare no conflict of interest.

References

1. Lee, J.C.; Dietz, P.; Leigh, D.; Yerazunis, W.; Hudson, S.E. Haptic pen: A tactile feedback stylus for touch screens. In Proceedings of the ACM Symposium on User Interface Software and Technology (UIST) 2004, Santa Fe, NM, USA, 24–27 October 2004; pp. 291–294. [CrossRef]
2. Kyung, K.; Lee, J.; Park, J. Ubi-Pen: Development of a compact tactile display module and its application to a haptic stylus. In Proceedings of the Second Joint EuroHaptics Conference and Symposium on Haptic Interfaces for Virtual Environment and Teleoperator Systems (WHC'07), Tsukaba, Japan, 22–24 March 2007; pp. 109–114. [CrossRef]
3. Kyung, K.; Lee, J.; Park, J. Haptic stylus and empirical studies on braille, button, and texture display. *J. Biomed. Biotechnol.* **2008**, *2008*, 369651. [CrossRef] [PubMed]
4. Liao, C.; Guimbretière, F.; Loeckenhoff, C.E. Pen-top feedback for paper-based interfaces. In Proceedings of the ACM Symposium on User Interface Software and Technology (UIST), New York, NY, USA, 15–18 October 2006; pp. 201–210. [CrossRef]
5. Cho, Y.; Bianchi, A.; Marquardt, N.; Bianchi-Berthouze, N. RealPen: Providing realism in handwriting tasks on touch surfaces using auditory-tactile feedback. In Proceedings of the ACM Symposium on User Interface Software and Technology (UIST), Tokyo, Japan, 16–19 October 2016; pp. 195–205. [CrossRef]
6. Wang, Q.; Ren, X.; Sun, X. Enhancing pen-based interaction using electrovibration and vibration haptic feedback. In Proceedings of the ACM CHI Conference on Human Factors in Computing Systems, New York, NY, USA, 6–11 May 2017; pp. 3746–3750. [CrossRef]
7. Arasan, A.; Basdogan, C.; Sezgin, T.M. Haptic stylus with inertial and vibro-tactile feedback. In Proceedings of the IEEE World Haptics Conference, Daejeon, Korea, 14–17 April 2013; pp. 425–430. [CrossRef]
8. Kamuro, S.; Minamizawa, K.; Kawakami, N.; Tachi, S. Ungrounded kinesthetic pen for haptic interaction with virtual environments. In Proceedings of the IEEE International Symposium on Robot and Human Interactive Communication, Toyama, Japan, 27 September–2 October 2009; pp. 436–441. [CrossRef]
9. Tian, L.; Song, A.; Chen, D. A novel haptic stylus for mobile terminal. In Proceedings of the EuroHaptics, London, UK, 4–7 July 2016; pp. 338–349. [CrossRef]
10. Kianzad, S.; MacLean, K.E. Collaborating through magic pens: Grounded forces in large, overlappable workspaces. In Proceedings of the AsiaHaptics, Incheon, Korea, 14–16 November 2018; pp. 233–237. [CrossRef]
11. Kara, O.C.; Patoglu, V. VnStylus: A haptic stylus with variable tip compliance. *IEEE Trans. Haptics* **2020**. Early Access. [CrossRef] [PubMed]
12. Chen, D.; Song, A.; Tian, L. A novel miniature multi-mode haptic pen for image interaction on mobile terminal. In Proceedings of the IEEE International Symposium on Interaction on Mobile Terminal Haptic, Audio and Visual Environments and Games (HAVE), Ottawa, ON, Canada, 11 October 2015; pp. 1–6. [CrossRef]
13. Chen, D.; Song, A.; Tian, L.; Yu, Y.; Zhu, L. MH-Pen: A pen-type multi-mode haptic interface for touch screens interaction. *IEEE Trans. Haptics* **2018**, *11*, 555–567. [CrossRef] [PubMed]
14. Carlson, J.D.; Jolly, M.R. MR fluid, foam and elastomer devices. *Mechatronics* **2000**, *10*, 555–569. [CrossRef]
15. Boelter, R.; Janocha, H. Design rules for MR fluid actuators in different working modes. In Proceedings of the Volume 3045, Smart Structures and Materials 1997: Passive Damping and Isolation, San Diego, CA, USA, 3–6 March 1997; pp. 148–159. [CrossRef]
16. Yang, T.-H.; Koo, J.H.; Kim, S.-Y.; Kyung, K.-U.; Kwon, D.-S. A miniature magneto-rheological actuator with an impedance sensing mechanism for haptic applications. *J. Intell. Mater. Syst. Struct.* **2013**, *25*, 1054–1061. [CrossRef]

17. Yang, T.-H.; Koo, J.H. Experimental evaluation of a miniature MR device for a wide range of human perceivable haptic sensations. *Smart. Mater. Struct.* **2017**, *26*, 125006. [CrossRef]
18. Brigley, M.; Choi, Y.-T.; Wereley, N.M.; Choi, S.-B. Magnetorheological isolators using multiple fluid modes. *J. Intell. Mater. Syst. Struct.* **2007**, *18*, 1143–1148. [CrossRef]
19. Grunwald, A.; Olabi, A.G. Design of magneto-rheological (MR) valve. *Sens. Actuators A Phys.* **2008**, *148*, 211–223. [CrossRef]
20. Jones, L.A. Kinesthetic sensing. In *Human and Machine Haptics*; MIT Press: Cambridge, MA, USA, 2000.
21. Immersion, UX Impacts of Haptic Latency in Automotive Interfaces. Available online: https://pdfs.semanticscholar.org/de7c/0c19188642cc925c256608f048163ea5973d.pdf (accessed on 28 August 2019).
22. GNOME. GNOME Human Interface Guidelines 2.2.3/ Feedback. Available online: https://developer.gnome.org/hig-book/unstable/feedback-response-times.html.en (accessed on 28 August 2019).
23. Nielsen Norman Group. Response Times: The 3 Important Limits. Available online: https://www.nngroup.com/articles/response-times-3-important-limits/ (accessed on 28 August 2019).

actuators

MDPI

Article

Piezoelectric Actuators for Tactile and Elasticity Sensing

Javier Toledo *, Víctor Ruiz-Díez, Jorge Hernando-García and José Luis Sánchez-Rojas

Microsystems, Actuators and Sensors Lab, Universidad de Castilla-La Mancha, 13071 Ciudad Real, Spain; victor.ruiz@uclm.es (V.R.-D.); jorge.hernando@uclm.es (J.H.-G.); joseluis.saldavero@uclm.es (J.L.S.-R.)

* Correspondence: javier.toledo@uclm.es; Tel.: +34-926-295-300 (ext. 96667)

Received: 10 February 2020; Accepted: 18 March 2020; Published: 19 March 2020

Abstract: Piezoelectric actuators have achieved remarkable progress in many fields, being able to generate forces or displacements to perform scanning, tuning, manipulating, tactile sensing or delivering functions. In this work, two piezoelectric PZT (lead zirconate titanate) bimorph actuators, with different tip contact materials, were applied as tactile sensors to estimate the modulus of elasticity, or Young's modulus, of low-stiffness materials. The actuators were chosen to work in resonance, taking advantage of a relatively low resonant frequency of the out-of-plane vibrational modes, associated with a convenient compliance, proven by optical and electrical characterization. Optical measurements performed with a scanning laser vibrometer confirmed that the displacement per applied voltage was around 437 nm/V for the resonator with the lower mass tip. In order to determine the modulus of elasticity of the sensed materials, the stiffness coefficient of the resonator was first calibrated against a force sensor, obtaining a value of 1565 ± 138 N/m. The actuators were mounted in a positioning stage to allow approximation and contact of the sensor tip with a set of target materials. Electrical measurements were performed using the resonator as part of an oscillator circuit, and the modulus of elasticity of the sample was derived from the contact resonant frequency curve of the cantilever–sample system. The resulting sensor is an effective, low-cost and non-destructive solution compared to atomic force microscopy (AFM) techniques. Materials with different modulus of elasticity were tested and the results compared to values reported in the literature.

Keywords: piezoelectric; PZT; actuator; out-of-plane; low-cost; tactile; modulus of elasticity; sensing

1. Introduction

Creating devices able to operate at the micrometre scale has been part of the scope of the vanguard of science and technology for many years. Using the same technology that has allowed the miniaturization of electronic circuits, it is possible to fabricate miniaturized systems composed of mechanical structures and electronic components, the so-called microelectromechanical systems (MEMS).

In recent years, sensors and actuators have been usually driven by electrostatic [1], electromagnetic [2], thermal [3], piezoelectric [4] and Lorentz forces [5,6]. The use of piezoelectric excitation is challenging, since a considerable voltage is needed to achieve practical forces or displacements. Nevertheless, in-plane piezoelectric actuators, using PZT (lead zirconate titanate) as a piezoelectric layer, have recently shown promising results: laser-machined [7] and thick-film PZT-based actuators [8] in the millimetre size range demonstrated displacements of 60 and 12 nm/V, while sub-millimetre-sized thin-film PZT-based actuators reached in-plane displacements as high as 300 nm/V [9–11].

Actuators, in general, have achieved remarkable progress in many fields [12,13], being able to generate forces or displacements to perform scanning, tuning, manipulating, tactile sensing or delivering functions [14,15]. Tactile sensing might be one of the most complex sensing modalities compared to sight, hearing, smell, and taste, as it is not a simple transduction of one physical property into a bioelectric signal [16]. In this field, contact resonance is a well-established technique, which allows for the determination of sample mechanical properties by tracking the resonance frequency of a structure modal vibration while interacting with the sample. The background of this technique was established by Kleesattel and Gladwell in 1968 [17–20]. Afterwards, Omata et al. used it at the micrometre level [21,22]; and Rabe deepened the understanding of the technique for atomic microscopy systems with publications that are a reference in the field [23,24]. In this case, a micro cantilever with a tip is used to sense information about the sample, including the force, roughness, and modulus of elasticity [25,26]. Nevertheless, in atomic force microscopy (AFM), the interaction force between the tip and the sample is obtained by detecting the deformation of the cantilever using a laser diode [27,28].

Over the last years, several works have been published related to this topic. For example, Fu et al. described a method for determining the modulus of elasticity using a sensor made of a piezoelectric bimorph cantilever, with a strain gauge for tactile sensing [29]. More recently, Bertke et al. described a piezo-resistive silicon cantilever, with a phase-locked loop (PLL) system, for controlled micro-tactile measurements based on contact resonance spectroscopy [30].

In this work, a piezoelectric PZT-based beam actuator was used as a tactile sensor in order to determine the modulus of elasticity of different materials with the help of low-cost driving electronic circuits. In this case, the actuator was first calibrated against a force sensor and mounted in a positioning stage to allow the approximation and the contact of the sensor tip with a set of target materials. Simultaneously, electrical measurements were performed using the resonator as part of a low-cost oscillator circuit, and the modulus of elasticity of the sample was derived from the contact resonant frequency curve of the cantilever–sample system. This method offers an effective, non-destructive and low-cost solution for tactile sensing compared to atomic force microscopy (AFM) techniques.

2. Materials and Methods

In this section, the bimorph actuators and their out-of-plane vibration modes, measurement setups and electronic circuits are described.

2.1. Lead Zirconate Titanate (PZT) Actuator

The PZT bimorph actuator used in this work is a low-cost and commercially available device of $15 \times 1.5 \times 0.6$ mm^3 (see Figure 1) [31]. It was initially characterized with an impedance analyzer and a scanning laser vibrometer (Polytec MSV400) to detect the different out-of-plane vibration modes and to obtain the main resonance parameters, resonant frequency (f_r) and quality factor (Q) of each vibration mode. Following Leissa's nomenclature, the vibration modes were designated as 10-mode and 20-mode according to the number of nodal lines in each direction [32].

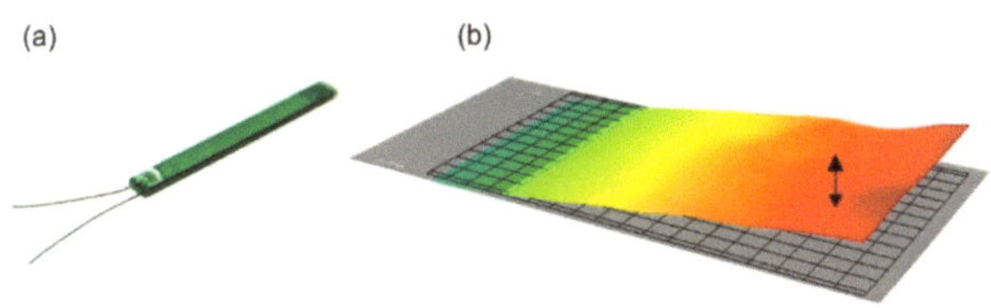

Figure 1. (**a**) Micrograph of the (lead zirconate titanate) PZT actuator, (**b**) optically measured modal shape (20-mode) with a laser Doppler vibrometer.

The results from impedance measurements are shown in Table 1. As it can be observed, the measured quality factor in air is in the range of 15 to 20 for these vibration modes, this being a relatively low value compared to other state-of-the-art resonators [33]. Nevertheless, the high conductance peak value (ΔG) along with the low resonant frequency of the 20-mode makes it suitable for the inclusion of the piezoelectric resonator in an oscillator circuit for a number of applications.

Table 1. Measurement of the resonant frequency, quality factor and conductance of the free-clamped bimorph actuator for different vibration modes with an impedance analyzer.

Vibration Mode	f_r [Hz]	Q	ΔG [μS]
10-mode	1191	15.85	0.954
20-mode	7190	20.27	20.2

Once the 20 mode was selected as the appropriate vibration mode, it was analyzed with a scanning laser vibrometer for different excitation voltages in order to measure the maximum displacement of the actuator versus applied voltage (see Table 2). As it can be observed, a linear relationship between the voltage and the displacement was obtained. As expected, the maximum displacement was measured in the resonator tip, which was used to contact the target material samples in our experiments. In order to check the influence of the tip on the modulus-of-elasticity sensing, two PZT actuators with different tips were employed in the measurements. In the named PZT-1 actuator, an aluminum tip was glued to the beam, while in the PZT-2 actuator a tungsten tip was used (see Figure 2).

Table 2. Measured resonant frequency (f_r) and displacement for actuators with different tips.

Device	Vibration Mode	f_r [kHz]	Voltage [V]	Displacement [nm]
PZT-1	20-mode	11	0.1	38.6
			0.5	180
			1	350
PZT-2	20-mode	11.5	0.1	40.8
			0.5	221
			1	437

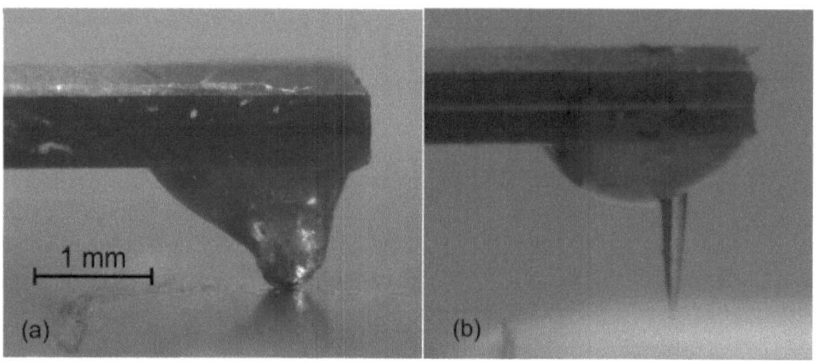

Figure 2. Picture of the different tips fixed to the PZT actuator. (**a**) PZT-1 actuator: aluminium tip; (**b**) PZT-2 actuator: tungsten tip.

In the results presented in Table 1, the beam boundary condition was clamped-free. Besides, in order to perform the elastic constant measurements, it was necessary to design a framework that implemented the clamped condition of the cantilever at one end. This framework, in which the beam's end was introduced and glued with epoxy, was fabricated with a FDM (Fused Deposition Modeling) 3D printer to minimize possible vibrations. Once the bimorph actuator and tip were glued, an increase of the resonant frequency was observed as it can be seen in Table 2, compared to Table 1 for the 20-mode. This occurs because the effective length of the cantilever is reduced when introduced and bonded inside the 3D-printed frame. As it can be observed, the PZT-2 actuator presents a higher displacement per voltage applied compared to the PZT-1 actuator, as expected when the size and mass of the tip are lower in the former.

2.2. Measurement Setup

In order to perform the modulus of elasticity measurements and allow the approximation and contact of the cantilever tip, a positioning stage was implemented as displayed in Figure 3. In this setup, the 3D-printed framework was attached to a platform controlled by six stepping motors that allow movement in any direction: three high-precision piezoelectric motors with a maximum displacement of 20 μm and three stepper motors that allow a coarse displacement of 5 mm. The approximation of the PZT actuator to the sample was performed with the help of the stepper motors, while the piezoelectric motors' high-precision positioning feedback was used for the determination of the modulus of elasticity. In order to control the motors at the positioning stage and simultaneously obtain the resonance parameters, a virtual instrument (VI) was designed in *LabView* from National Instruments [34].

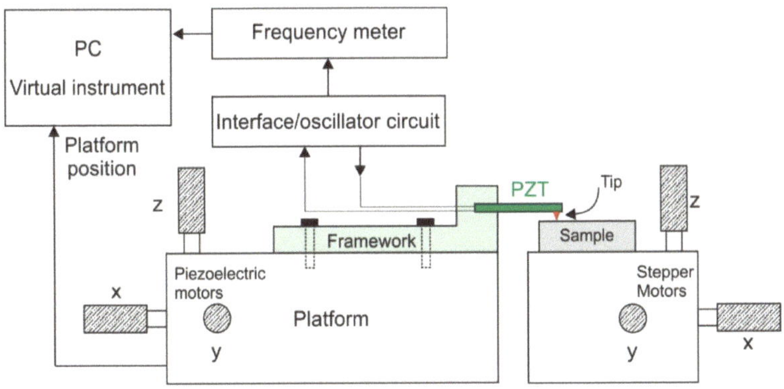

Figure 3. Schematic of the setup for the modulus of elasticity measurements.

2.3. Low-Cost Electronic Circuits

In this section, the designed driving interface and oscillator circuits, for the open and closed-loop measurements described in Section 3, are introduced.

2.3.1. Interface Circuit

Firstly, the behaviour of the resonator was tested by making open-loop measurements. These were performed using a lock in amplifier to excite and collect the signal from the actuator within an interface circuit (see Figure 4) [35]. The main objective of the interface circuit is to reduce the parasitic effects of the resonator and to obtain an appropriate resonance curve for the later closed-loop measurements. Both, the PZT actuator and a parasitic compensating device (PZT compensation) were used simultaneously for actuation and sensing [36].

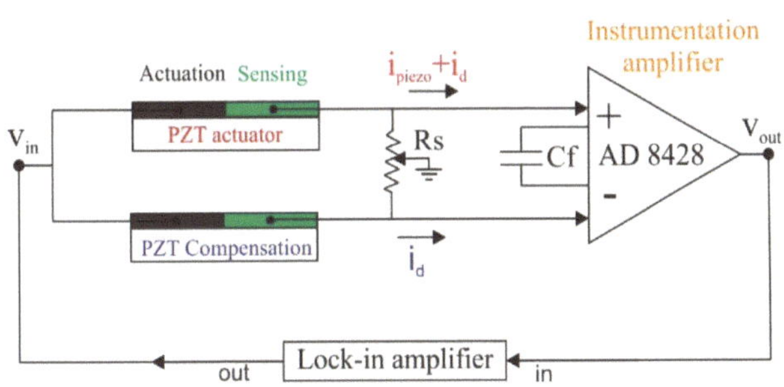

Figure 4. Schematic of the interface circuit designed for the open-loop measurements.

The designed interface circuit uses an instrumentation amplifier (AD8428) to subtract the compensating device dielectric current (i_d) from that of the actuator (i_{piezo} + i_d), obtaining a voltage proportional to the piezoelectric current (i_{piezo}), which is proportional to the vibration amplitude, at the output of the instrumentation amplifier (V_{out}).

As it can be observed in Figure 4, the compensation can be accurately tuned by implementing variable resistance (Rs) at the input of the instrumentation amplifier. The compensating device (PZT compensation) is nominally identical to the PZT actuator, but clamped all along its length, to prevent vibrations. Since the materials and dimensions in the compensating and actuator devices are nominally the same, they are expected to show similar electrical behavior with respect to their parasitic effects.

2.3.2. Oscillator Circuit

Once the parasitic effects of the resonator were minimized with the previous interface circuit, we included the actuator in an oscillator circuit scheme. The possibility of reducing the total size, cost and power consumption of the system makes the oscillator circuit a very interesting solution for any potential application scenario [37,38]. In addition, the reasonably good value of the admittance peak amplitude, along with the low natural frequency of the actuator, allowed for a great flexibility and simplicity in the design of the oscillator circuit.

In this case, the interface circuit was modified to meet the Barkhausen criterion when closing the loop between V_{out} and V_{in}, and the lock-in amplifier was removed. In order to achieve this, a higher gain in the instrumentation amplifier was applied, replacing the resistance that controls the gain amplifier by a capacitor (C_f) (see Figure 4). This component acts as a low-pass filter and introduces the gain necessary to meet the Barkhausen criterion. Furthermore, the phase shift when closing the loop is 0°, since the instrumentation amplifier does not introduce any phase shift. This means that the oscillation frequency and the real resonance frequency of the actuator are equal.

Finally, when a stable oscillation is attained at the output (V_{out}), the frequency value can be easily tracked by a frequency counter (Agilent 53220a). This is controlled by the same virtual instrument that drives the positioning platform. In this case, 1000 samples were obtained in the 20 μm range of the piezoelectric motors, with a sample rate of 0.8 s.

3. Measurements and Results

3.1. Actuator Measurements

The analysis of the different samples was performed with the previously described low cost electronic circuits in open and closed-loop schemes.

3.1.1. Open-Loop Measurements

Firstly, the behavior of the interface circuit (see Figure 4) was tested in an open-loop configuration with the PZT-1 actuator. In order to perform these measurements, the platform and actuator were moved with the virtual instrument to approach the tip and identify the contact on a sample of polylactic acid material, commonly known as PLA (see Table 3), frequently used in 3D printing. In this case, the lock-in amplifier performed the frequency sweep measurement during the contact approaching process. Once the actuator was brought into contact with the PLA sample, three open-loop measurements were completed by moving the platform between the positions 7.15 and 7.75 μm in the z-axis direction.

Table 3. Modulus of elasticity reference values of the different materials analyzed.

Material	E [GPa]
PDMS (Polydimethylsiloxane)	0.0005–0.0037 [39]
Rubber	0.001–0.1 [40]
	1.4–3.1 [40]
ABS (Acrylonitrile Butadiene Styrene)	1.79–3.2 [41]
	2.1 [42]
Nylon (Synthetic polymers based on polyamides)	2–4 [40]
PLA (Polylactic acid)	2.02–3.55 [43–45]
	3.5 [42]
Aluminum	69 [40,46]

As it can be seen in Figure 5, a higher force or displacement leads to an increase of the resonant frequency, indicating that the procedure is valid for the elastic-constant-sensing application. Although these open-loop measurements were only used as a validation step, it was also checked that the phase curve met the Barkhausen criterion for the oscillation of the resonator to start, a necessary requirement for the later implementation of the oscillator circuit.

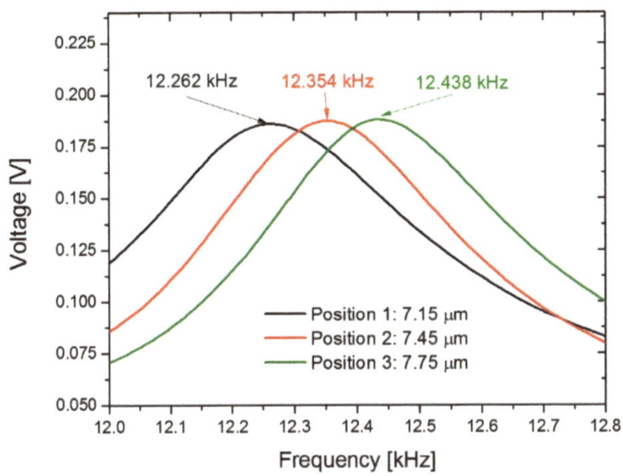

Figure 5. Open-loop measurements of the PZT-1 actuator in contact with PLA sample with a displacement of 0.3 μm between measurements.

3.1.2. Closed-Loop Measurements

Once the behavior of the actuator and the measurement setup was verified, different closed-loop measurements were performed using the previously described oscillator circuit (see Section 2.3.2). In Table 3, the different materials are shown from lowest to highest modulus of elasticity and their reference values: PDMS, rubber, ABS, nylon, PLA and aluminum. As it can be seen, the PDMS and the rubber present a similar elastic constant. The same occurs for the ABS, Nylon and PLA samples. Nevertheless, it is possible to observe differences in the resonant frequency values for both resonators (see Figures 6 and 7).

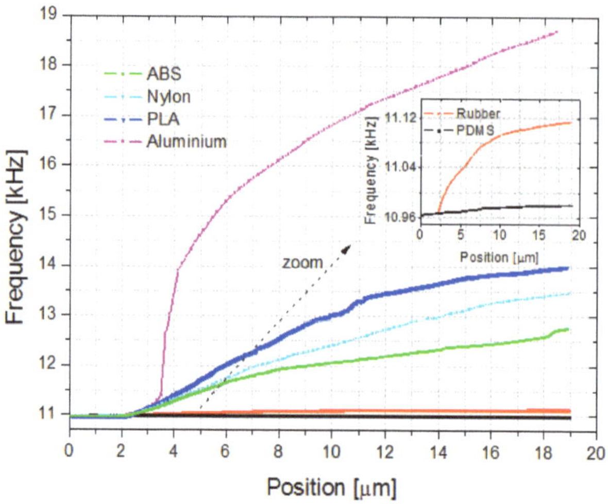

Figure 6. Closed-loop frequency measurements for the PZT-1 actuator.

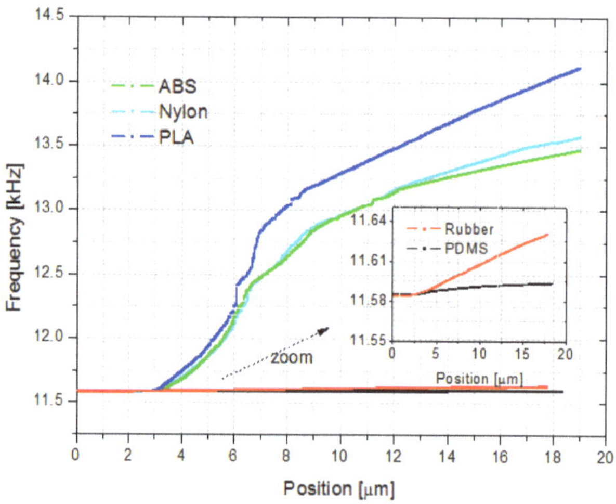

Figure 7. Closed-loop frequency measurements for the PZT-2 actuator.

As it can be seen in Figures 6 and 7, a higher force or displacement increases the resonant frequency, as was also verified with the open-loop measurements. In addition, the resonant frequency changes accordingly with the modulus of elasticity of the sample, indicating that the resonant frequency could be used as a valid parameter for the determination of the modulus of elasticity. As the results suggest, it would be necessary to implement a force control capable of monitoring the applied force on the sample surface. Nevertheless, in this work, an alternative solution was tested by taking as a reference the point where the tip makes contact with the target sample, and therefore the resonant frequency increases. This allowed us to compare the frequency measurements of different materials and also to estimate the applied force through the actuator calibration process described in Section 3.2.

The procedure performed to obtain the modulus of elasticity of the samples can be summarized as follows. Firstly, the resonator is excited and brought into contact with the sample with a tip-target approaching process. After that, the resonator tip contacts the sample, the applied force is obtained from the position of the platform (see Section 3.2) and the resonant frequency is tracked with the oscillator circuit and the frequency counter.

As it can be observed in the closed-loop measurements performed with the PZT-2 actuator, almost no difference in the resonant frequency value, during the initial displacement of approximately 4 μm of the positioning stage, is detected (see Figure 7). This is due to the fact that the frequency shift observed during this displacement, which is similar for the different materials, may be attributed to the elasticity of the epoxy glue used to attach the tungsten tip to the actuator. Once the initial range is exceeded, the tip remains fixed to the actuator, being able to detect differences in its resonant frequency as the modulus of elasticity of the tungsten tip (450 GPa) is much greater than the modulus of elasticity of the actuator (33.3 GPa).

As it can be seen in Figures 6 and 7, a higher resonant frequency was obtained for PLA, due to its higher elastic constant compared to ABS and nylon. In contrast, almost no differences were observed for PDMS and rubber, due to the low modulus of elasticity of these materials compared with the actuators and the tips. These results confirm the results of previous studies on AFM, where a high detection sensitivity, or stiffer or flexible cantilevers, are required for testing on samples with high or low modulus of elasticity, respectively [47].

3.2. Actuator and Force Sensor Calibration

The main objective of this work was to demonstrate a simple method to obtain the modulus of elasticity of different materials, applying it to those typically used in 3D printing technology. For this reason, it is necessary to perform two different approaches in order to correlate the contact resonant frequency of the actuator with the applied force on the surface material, and its elastic properties.

Firstly, the stiffness constant (K) of the PZT-2 actuator was measured, since this parameter is required for the determination of the modulus of elasticity (see Section 3.3). In this case, a commercial force sensor (see Figure 8) [48] allowed us to obtain a linear relationship between its output voltage and the force applied by the piezoelectric beam when the positioning stage was shifted (see Figure 9). According to the datasheet of the force sensor, this presents a sensitivity of 7.2 mV/V/N; since the force sensor is biased with a voltage of 10 V, we obtain a conversion factor of 72 mV/N. With this conversion factor we can finally obtain the stiffness constant (K) of the actuator. An average value of K, and the deviation obtained with five different measurements, leads to 1565 ± 138 N/m, similar to the value of 1520 N/m reported in the datasheet.

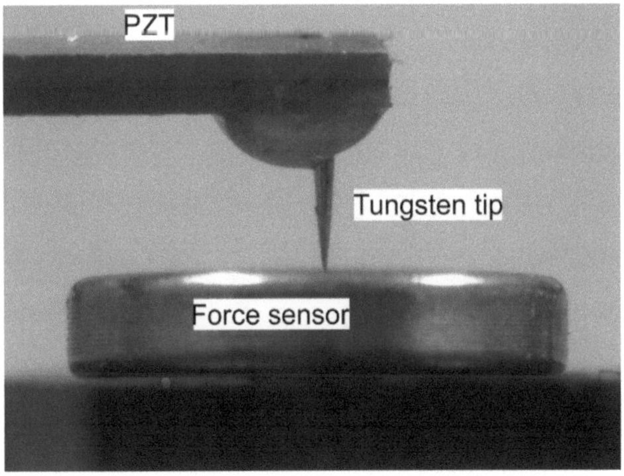

Figure 8. Calibration of the PZT-2 actuator with a force sensor.

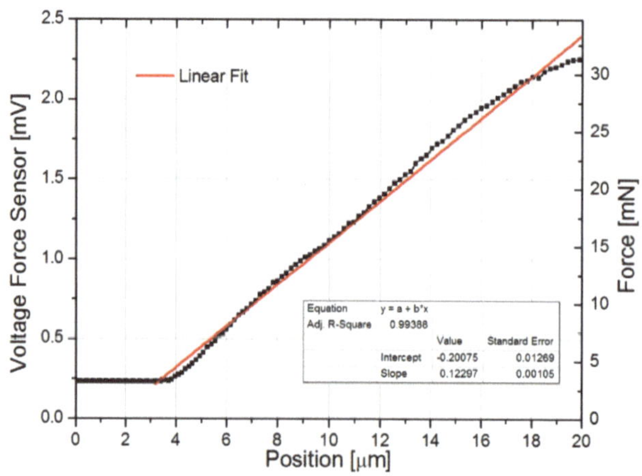

Figure 9. Calibration of the stiffness constant for the PZT-2 actuator with the force sensor.

Once the stiffness contact is known, it is necessary to obtain the force applied by the actuator when the tip makes contact with the material to be analyzed. To do this, the actuator is excited with the electronics and brought into contact with the force sensor. In this case, when the resonance curve changes, indicating that the actuator has made contact with the sample, the position of the platform is taken as a reference. This approach allows us to estimate the force applied by the actuator over the sample just by sampling the position of the platform.

3.3. *Modulus of Elasticity Sensing*

In this section, the performance of the tactile sensor for the determination of the modulus of elasticity is carefully examined. Once the calibration of the PZT-2 actuator was performed, the modulus of elasticity of the different materials were obtained using the Hertz contact theory [49,50], following a procedure previously reported [29].

In a first step, the modulus of elasticity of the PZT-2 actuator (E) and the moment of inertia (I) were calculated through Equations (1) and (2).

$$E = \frac{4Kl^3}{wt^3} \tag{1}$$

$$I = \frac{wt^3}{12} \tag{2}$$

where K is the stiffness constant previously obtained, with a value of 1565 N/m. In this case, the length of the cantilever beam is shorter due to the anchoring to the 3D-printed framework, being now of length (l) 12 mm, width (w) 1.5 mm and thickness (t) 0.6 mm. Using this data, we could obtain a modulus of elasticity (E) for the PZT actuator of 33.3 GPa.

In order to estimate the modulus of elasticity of the samples (E_s in Equation (5)), it is necessary to first calculate the constants C_c and k_t defined in Equations (3) and (4). Once the modulus of elasticity of the actuator was known, and taking into account that the density of the PZT actuator (ρ) is 7500 kg/m^3, it was possible to estimate the constant C_c, obtaining a value of 0.0498 s$^{1/2}$ (see Equation (3)).

The value of the constant C_c presents different error sources, such as the dimensions and properties of the cantilever. In order to reduce the final error, the constant C_c was calibrated through the following process. In this case, four different positions of the platform (7, 11, 15 and 18 μm) and the corresponding values of resonant frequency, force and modulus of elasticity for the ABS, nylon and PLA, were taken references in Equations (4) and (5) (see Table 4).

$$C_c = l\ \sqrt{2\pi}\sqrt[4]{\frac{\rho A}{EI}} \tag{3}$$

$$k_t = \frac{K\left(C_c\sqrt{f_r}\right)^3}{3}\frac{\cos\left(C_c\sqrt{f_r}\right)\cosh\left(C_c\sqrt{f_r}\right)+1}{\sinh\left(C_c\sqrt{f_r}\right)\ \cos\left(C_c\sqrt{f_r}\right)-\sin\left(C_c\sqrt{f_r}\right)\ \cos\left(C_c\sqrt{f_r}\right)} \tag{4}$$

$$E_s = \frac{\left(1-v_s^2\right)E_t\sqrt{k_t^3/6RF}}{E_t-\left(1-v_s^2\right)\sqrt{k_t^3/6RF}} \tag{5}$$

Table 4. Materials analysed with the PZT-2 actuator and their reference values published in the literature.

Material	Position [μm]	Force [mN]	Frequency [Hz]	$E_{s\text{-Estimated}}$ [GPa]	$E_{s\text{-Reference}}$ [GPa]
ABS	7	10	12,465	2.73 ± 0.129	1.4–3.1 [40]
	11	17.5	13,038		1.79–3.2 [41]
	15	24	13,302		2.1 [42]
	18	30	13,430		2.03 [51]
					1.15–1.96 [52]
Nylon	7	10	12,474	2.79 ± 0.077	2–4 [40]
	11	17.5	13,040		
	15	24	13,364		
	18	30	13,528		
PLA	7	10	12,828	3.5 ± 0.022	2.02–3.55 [43–45]
	11	17.5	13,381		
	15	24	13,778		3.5 [42]
	18	30	14,034		

In Equations (4) and (5), E_t represents the modulus of elasticity of the tungsten tip with a value of 450 GPa, E_s is the modulus of elasticity of the measured sample, R is the radius of the tip with a value of 10 μm, v_s is the Poisson's coefficient with a value of 0.35, k_t is the tip stiffness constant, K is the cantilever stiffness constant, f_r is the contact measured resonant frequency and F is the applied force. With all these data, and using Equations (4) and (5), it was possible to calibrate the constant C_c obtaining a new value of 0.0458 $s^{1/2}$. Once this constant C_c is known, the PZT actuator can be used to estimate the modulus of elasticity of different materials. Table 4 displays a summary with the estimated and reference values for the modulus of elasticity.

4. Discussion

In this work, the modulus of elasticity, or Young's modulus, of different materials was derived using a PZT actuator, a low-cost oscillator circuit and a positioning stage. Other previous publications suggest it is necessary to implement a force control or gauge deformation capable of monitoring the applied force and deformation of the actuator [27,29]. Nevertheless, in this work, an alternative solution was tested, by taking the point where the PZT actuator tip makes contact with the sample, and therefore the resonant frequency increases, as a reference, allowing us to compare the frequency measurements of different materials and also to estimate the applied force.

Another advantage of this work is that the designed setup, positioning stage and electronics could be easily applied to different actuators with out-of-plane vibration modes. In this case it would only be necessary to design a new 3D-printed framework to fix the actuator and to calibrate the actuator against the force sensor to obtain its stiffness constant. This approach would also make it possible to extend the application to different actuators, in order to analyze materials with different modulus of elasticity. Also noteworthy is the novelty of the electronics circuits implemented in this work, making it possible to reduce parasitic effects and to obtain an appropriate resonance curve for the oscillation of the piezoelectric actuator.

The procedure implemented to obtain the modulus of elasticity of the samples can be summarized as follows. First, the actuator, with a tip attached to its end, is excited near resonance and brought into contact with the sample while the resonance is assessed. Once the resonance curve indicates that the tip contacts the sample, the position of the platform is taken, and the resonant frequency is tracked with the oscillator circuit and the frequency counter.

As indicated in previous studies on AFM, to get a high detection sensitivity, stiff or flexible actuators are required for testing on samples with a high or low modulus of elasticity, respectively [47]. This fact was also observed in our results. For example, a higher resonant frequency was obtained for the PLA due to its higher elastic constant. On the contrary, almost no differences were observed for the PDMS and rubber due to the low modulus of elasticity of these materials compared with the PZT actuator.

In this work, the modulus of elasticity was obtained through the Hertz contact theory, obtaining a resolution, through the measured frequency values for different positions of the platform, of 0.129 GPa in the worst case. As it can be observed in Table 4, the estimated modulus of elasticity is in the range of the reference values published in previous work [40–45,51,52]. Nevertheless, the deviation of the modulus of elasticity for the ABS may be higher if compared with the average of published reference values. There are several reasons that could explain this behavior. The reference values cannot be fairly compared with our estimated values since the method developed in our work is based on a dynamic procedure limited to a reduced area and deformation, and not to the whole sample [53]. This conclusion has also been observed in other works where large deviations were obtained in the estimation of the modulus of elasticity between the ABS filament itself and the 3D-printed sample [42], where the data obtained from tensile tests of the filaments were also only qualitatively consistent with those provided by the manufacturer. Another reason behind the deviations with the reference values is the lack of standard test methods for the determination of the mechanical properties of parts manufactured using FDM. For example, the standards ASTM D638 [54] and D790 [53] may be applied for testing tensile and flexural specimens, respectively [43].

Although one of the goals of this work was to estimate the Young's modulus of the different samples, further investigation could be necessary to check the deviation with different measurement methods, and the viability of the piezoelectric actuator with different materials and setup.

5. Conclusions

In conclusion, a PZT bimorph actuator was tested for the determination of the modulus of elasticity, using low-cost electronic circuits based on an interface and oscillator circuit, and obtaining an accurate performance. For this reason, the PZT actuator along with the positioning stage can be considered an effective and non-destructive solution for the determination of the modulus of elasticity and sample characterization. In this case, the estimates of the modulus of elasticity of ABS, nylon and PLA were in the range of the reference values published in previous works, and a resolution of 0.129 GPa was obtained in the worst case. Furthermore, sample information such as the position, orientation, and surface topography could also be obtained by adding scanning functionality to the testing platform.

Author Contributions: Investigation, J.T. and J.L.S.-R; Project administration, J.L.S.-R.; Supervision, V.R.-D. and J.H.-G.; Validation, J.T.; Writing—original draft, J.T.; Writing—review & editing, J.T., V.R.-D., J.H.-G. and J.L.S.-R. All authors have read and agreed to the published version of the manuscript.

Funding: This research was funded by the European Regional Development Funds through projects RETOS RTI2018-094960-B-I00 and regional JCCLM SBPLY/17/180501/000139.

Conflicts of Interest: The authors declare no conflict of interest. The funders had no role in the design of the study; in the collection, analyses, or interpretation of data; in the writing of the manuscript, or in the decision to publish the results.

References

1. Grade, J.D.; Jerman, H.; Kenny, T.W. Design of Large Deflection Electrostatic Actuators. *J. Microelectromech. Syst.* **2001**, *XX*, 1–10. [CrossRef]
2. Ko, J.S.; Lee, M.L. Development and application of laterally driven electromagnetic microactuator. *Appl. Phys. Lett.* **2002**, *81*, 547–549. [CrossRef]
3. Qui, J.; Lang, J.H.; Slocum, A.H.; Strümpler, R. A high-current electrothermal bistable MEMS relay. In Proceedings of the Sixteenth Annual International Conference on Micro Electro Mechanical Systems, Kyoto, Japan, 23 January 2003; pp. 64–67.
4. Manzaneque, T.; Ruiz-Díez, V.; Hernando-García, J.; Ababneh, A.; Al-Omari, A.N.; Kucera, M.; Bittner, A.; Schmid, U.; Seidel, H.; Sánchez-Rojas, J.L. Piezoelectric in-plane microplate resonators based on contour and flexure-actuated modes. *Microsyst. Technol.* **2014**, *20*, 691–699. [CrossRef]
5. Lee, F.Y.; Tang, T.L.; Fang, W. Development of a novel dual-axis large-displacement microstage using Lorentz force actuators and curved-beam springs. *Procedia Eng.* **2011**, *25*, 689–692. [CrossRef]
6. Lv, X.D.; Wei, W.W.; Mao, X.; Yang, J.L.; Yang, F.H. A linear low driving voltage mems actuator with large lateral stroke driven by Lorentz force. In Proceedings of the 2015 Transducers—2015 18th International Conference on Solid-State Sensors, Actuators and Microsystems (TRANSDUCERS), Anchorage, AK, USA, 21–25 June 2015; pp. 2117–2120.
7. Leang, K.K.; Fleming, A.J. High-speed serial-kinematic SPM scanner: Design and drive considerations. *Asian J. Control* **2009**, *11*, 144–153. [CrossRef]
8. Aktakka, E.E.; Peterson, R.L.; Najafi, K. A 3-DOF piezoelectric micro vibratory stage based on bulk-PZT/silicon crab-leg suspensions. In Proceedings of the 2013 IEEE 26th International Conference on Micro Electro Mechanical Systems (MEMS), Taipei, Taiwan, 20–24 January 2013; pp. 576–579.
9. Conway, N.J.; Kim, S.-G. Large Strain, Piezoelectric, in-Plane Micro-Actuator. In Proceedings of the 17th IEEE International Conference on Micro Electro Mechanical Systems. Maastricht MEMS 2004 Technical Digest, Maastricht, The Netherlands, 25–29 January 2004; pp. 454–457.
10. Oldham, K.R.; Pulskamp, J.S.; Polcawich, R.G.; Dubey, M. Thin-film PZT lateral actuators with extended stroke. *J. Microelectromech. Syst.* **2008**, *17*, 890–899. [CrossRef]
11. Yoshida, S.; Wang, N.; Kumano, M.; Kawai, Y.; Tanaka, S.; Esashi, M. Fabrication and characterization of laterally-driven piezoelectric bimorph MEMS actuator with sol-gel-based high-aspect-ratio PZT structure. In Proceedings of the 2013 IEEE 26th International Conference on Micro Electro Mechanical Systems (MEMS), Taipei, Taiwan, 20–24 January 2013.
12. Howe, R.T.; Muller, R.S.; Gabriel, K.J.; Trimmer, W.S.N. Silicon micromechanics: Sensors and actuators on a chip. *IEEE Spectr.* **1990**, *27*, 29–31. [CrossRef]
13. Fujita, H. Microactuators and micromachines. *Proc. IEEE* **1998**, *86*, 1721–1732. [CrossRef]
14. Panchawagh, H.V.; Faheem, F.F.; Herrmann, C.F.; Serrell, D.B.; Finch, D.S.; Mahajan, R.L. Packaging of in-plane thermal microactuators for BioMEMS applications. In Proceedings of the American Society of Mechanical Engineers, Micro-Electro Mechanical Systems Division, (Publications) MEMS, Oahu, Hawaii, 5 November 2005.
15. Eltagoury, Y.M.; Soliman, M.; Al-Otaibi, M.; Sabry, Y.M.; Sadek, M.; Khalil, D. In-plane comb-drive actuator with high frequency-displacement product for micro-optical bench applications. In Proceedings of the International Conference on Optical MEMS and Nanophotonics, Glasgow, UK, 17–21 August 2014.
16. Lee, M.H.; Nicholls, H.R. Review Article Tactile sensing for mechatronics: A state of the art survey. *Mechatronics* **1999**, *9*, 1–31. [CrossRef]
17. Kleesattel, C.; Gladwell, G.M.L. The contact-impedance meter 1. *Ultrasonics* **1968**, *6*, 175–180. [CrossRef]
18. Gladwell, G.M.L.; Kleesattel, C. The contact-impedance meter 2. *Ultrasonics* **1968**, *6*, 244–251. [CrossRef]
19. Kleesattel, C.; Gladwell, G.M.L. The contact-impedance meter 3. *Ultrasonics* **1969**, *7*, 57–62. [CrossRef]
20. Kleesattel, C. Contact impedance meter. Part IV. *Ultrasonics* **1970**, *8*, 39–48. [CrossRef]

21. Omata, S.; Terunuma, Y. New tactile sensor like the human hand and its applications. *Sens. Actuators A. Phys.* **1992**, *35*, 9–15. [CrossRef]
22. Murayama, Y.; Omata, S. Fabrication of micro tactile sensor for the measurement of micro-scale local elasticity. *Sens. Actuators A Phys.* **2004**, *109*, 202–207. [CrossRef]
23. Rabe, U.; Janser, K.; Arnold, W. Vibrations of free and surface-coupled atomic force microscope cantilevere: Theory and experiment. *Rev. Sci. Instrum.* **1996**, *67*, 3281–3293. [CrossRef]
24. Rabe, U.; Arnold, W. Acoustic microscopy by atomic force microscopy. *Appl. Phys. Lett.* **1994**, *64*, 1493. [CrossRef]
25. Zhou, X.; Fu, J.; Li, Y.; Li, F. Nanomechanical mapping of glass fiber reinforced polymer composites using atomic force acoustic microscopy. *J. Appl. Polym. Sci.* **2014**, *131*. [CrossRef]
26. Rabe, U.; Kopycinska, M.; Hirsekorn, S.; Muñoz-Saldaña, J.; Schneider, G.A.; Arnold, W. High-resolution characterization of piezoelectric ceramics by ultrasonic scanning force microscopy techniques. *J. Phys. D. Appl. Phys.* **2002**, *35*, 2621–2635. [CrossRef]
27. Monclus, M.A.; Young, T.J.; Di Maio, D. AFM indentation method used for elastic modulus characterization of interfaces and thin layers. *J. Mater. Sci.* **2010**, *45*, 3190–3197. [CrossRef]
28. Bhushan, B.; Koinkar, V.N. Nanoindentation hardness measurements using atomic force microscopy. *Appl. Phys. Lett.* **1994**, *64*, 1653–1655. [CrossRef]
29. Fu, J.; Li, F. A forefinger-like tactile sensor for elasticity sensing based on piezoelectric cantilevers. *Sens. Actuators A Phys.* **2015**, *234*, 351–358. [CrossRef]
30. Bertke, M.; Fahrbach, M.; Hamdana, G.; Xu, J.; Wasisto, H.S.; Peiner, E. Contact resonance spectroscopy for on-the-machine manufactory monitoring. *Sens. Actuators A Phys.* **2018**, *279*, 501–508. [CrossRef]
31. Rs Pro. Piezoelectric Ceramic Bimorph Element. Available online: https://docs-emea.rs-online.com/webdocs/1582/0900766b81582396.pdf (accessed on 2 February 2020).
32. Leissa, A.W. *Vibration of Plates*; Scientific and Technical Information Division, National Aeronautics and Space Administration: Washington, DC, USA, 1969.
33. Toledo, J.; Ruiz-díez, V.; Pfusterschmied, G.; Schmid, U.; Sánchez-rojas, J.L. Flow-through sensor based on piezoelectric MEMS resonator for the in-line monitoring of wine fermentation. *Sens. Actuators B. Chem.* **2017**, *254*, 291–298. [CrossRef]
34. LabView, National Instruments. Available online: https://www.ni.com/es-es/shop/labview.html (accessed on 2 February 2020).
35. Toledo, J.; Manzaneque, T.; Ruiz-Díez, V.; Kucera, M.; Pfusterschmied, G.; Wistrela, E.; Schmid, U.; Sánchez-Rojas, J.L. Piezoelectric resonators and oscillator circuit based on higher-order out-of-plane modes for density-viscosity measurements of liquids. *J. Micromech. Microeng.* **2016**, *26*, 084012. [CrossRef]
36. Kucera, M.; Hofbauer, F.; Wistrela, E.; Manzaneque, T.; Ruiz-Díez, V.; Sánchez-Rojas, J.L.; Bittner, A.; Schmid, U. Lock-in amplifier powered analogue Q-control circuit for self-actuated self-sensing piezoelectric MEMS resonators. *Microsyst. Technol.* **2014**, *20*, 615–625. [CrossRef]
37. Toledo, J.; Ruiz-Díez, V.; Bertke, M.; Wasisto, H.S.; Peiner, E.; Sánchez-Rojas, J.L. Piezoelectric MEMS resonators for cigarette particle detection. *Micromachines* **2019**, *10*, 145. [CrossRef]
38. Toledo, J.; Ruiz-Díez, V.; Pfusterschmied, G.; Schmid, U.; Sánchez-Rojas, J.L. Characterization of oscillator circuits for monitoring the density-viscosity of liquids by means of piezoelectric MEMS microresonators. In Proceedings of the Smart Sensors, Actuators, and MEMS VIII, Barcelona, Spain, 8–11 May 2017.
39. Wang, Z.; Volinsky, A.A.; Gallant, N.D. Crosslinking effect on polydimethylsiloxane elastic modulus measured by custom-built compression instrument. *J. Appl. Polym. Sci.* **2014**, 131. [CrossRef]
40. Engineering ToolBox. Young's Modulus—Tensile and Yield Strength for Common Materials. Available online: https://www.engineeringtoolbox.com/young-modulus-d_417.html (accessed on 29 January 2020).
41. Acrylonitrile Butadiene Styrene (ABS) and its Features. Available online: https://omnexus.specialchem.com/selection-guide/acrylonitrile-butadiene-styrene-abs-plastic (accessed on 3 March 2020).
42. Rodríguez-Panes, A.; Claver, J.; Camacho, A.M. The influence of manufacturing parameters on the mechanical behaviour of PLA and ABS pieces manufactured by FDM: A comparative analysis. *Materials* **2018**, *11*, 1333. [CrossRef]
43. Chacón, J.M.; Caminero, M.A.; García-Plaza, E.; Núñez, P.J. Additive manufacturing of PLA structures using fused deposition modelling: Effect of process parameters on mechanical properties and their optimal selection. *Mater. Des.* **2017**, *124*, 143–157. [CrossRef]

44. Kotlinski, J. Mechanical properties of commercial rapid prototyping materials. *Rapid Prototyp. J.* **2014**, *20*, 499–510. [CrossRef]
45. Lanzotti, A.; Grasso, M.; Staiano, G.; Martorelli, M. The impact of process parameters on mechanical properties of parts fabricated in PLA with an open-source 3-D printer. *Rapid Prototyp. J.* **2015**, *21*, 604–617. [CrossRef]
46. ASM International. Properties and Selection: Irons, Steels, and High-Performance Alloys. In *ASM Handbook*; ASM International: Materials Park, OH, USA, 1993.
47. Turner, J.A.; Wiehn, J.S. Sensitivity of flexural and torsional vibration modes of atomic force microscope cantilevers to surface stiffness variations. *Nanotechnology* **2001**, *12*, 322. [CrossRef]
48. FSG Series Force Sensors. Available online: https://sensing.honeywell.com/fsg005wnpb-force-sensors2 (accessed on 19 March 2020).
49. Johnson, K.L. One hundred years of hertz contact. *Proc. Inst. Mech. Eng.* **1982**, *196*, 363–378. [CrossRef]
50. MacHado, M.; Moreira, P.; Flores, P.; Lankarani, H.M. Compliant contact force models in multibody dynamics: Evolution of the Hertz contact theory. *Mech. Mach. Theory* **2012**, *53*, 99–121. [CrossRef]
51. De León, A.S.; Domínguez-Calvo, A.; Molina, S.I. Materials with enhanced adhesive properties based on acrylonitrile-butadiene-styrene (ABS)/thermoplastic polyurethane (TPU) blends for fused filament fabrication (FFF). *Mater. Des.* **2019**, *182*, 108044. [CrossRef]
52. Casavola, C.; Cazzato, A.; Moramarco, V.; Renna, G. Mechanical behaviour of ABS-Fused Filament Fabrication compounds under impact tensile loadings. *Materials* **2019**, *12*, 1295. [CrossRef]
53. ASTM International. *D 790—Standard Test Method for Flexural Properties of Unreinforced and Reinforced Plastics and Electrical Insulation Materials*; ASTM: West Conshohocken, PA, USA, 2015.
54. ASTM International. *D638—Standard Test Method for Tensile Properties of Plastics*; ASTM International: West Conshohocken, PA, USA, 2014. [CrossRef]

MDPI AG
Grosspeteranlage 5
4052 Basel
Switzerland
Tel.: +41 61 683 77 34

Actuators Editorial Office
E-mail: actuators@mdpi.com
www.mdpi.com/journal/actuators

www.ingramcontent.com/pod-product-compliance
Lightning Source LLC
Chambersburg PA
CBHW041343210926
51850CB00013B/2971
* 9 7 8 3 7 2 5 8 3 6 8 3 3 *